Lecture Notes
in Control and Information Sciences 394

Editors: M. Thoma, F. Allgöwer, M. Morari

Goele Pipeleers, Bram Demeulenaere,
Jan Swevers

Optimal Linear Controller
Design for Periodic Inputs

 Springer

Authors

Dr. Goele Pipeleers

Katholieke Universiteit Leuven
Department of Mechanical Engineering
Division PMA
Celestijnenlaan 300B
B-3001 Heverlee
Belgium
E-mail: goele.pipeleers@mech.kuleuven.be

Prof. Jan Swevers

Katholieke Universiteit Leuven
Department of Mechanical Engineering
Division PMA
Celestijnenlaan 300B
B-3001 Heverlee
Belgium
E-mail: jan.swevers@mech.kuleuven.be

Dr. Bram Demeulenaere

Atlas Copco Airpower NV
Airtec Division
Boomsesteenweg 957
B-2610 Wilrijk
Belgium
E-mail: bram.demeulenaere@be.atlascopco.com

ISBN 978-1-84882-974-9   ISBN 978-1-84882-975-6 (eBook)

DOI 10.1007/978-1-84882-975-6

Lecture Notes in Control and Information Sciences   ISSN 0170-8643

Library of Congress Control Number: 2009937153

Typeset & Cover Design: Scientific Publishing Services Pvt. Ltd., Chennai, India.

Printed in acid-free paper

5 4 3 2 1 0

springer.com

Preface

Periodic reference and disturbance signals are widespread in engineering practice, as every rotating machine and repeated process involves periodicity. Exploiting the periodic input characteristics in the controller design is indispensable to meet tight performance demands in spite of measurement noise, model inaccuracies...

This monograph proposes a general design methodology for linear controllers facing periodic inputs, which applies to all controller types reported in the literature. The proposed design methodology is able to reproduce and outperform major current design approaches, where this superior performance stems from the following properties: (i) uncertainty on the input period is explicitly accounted for; (ii) periodic performance is traded-off against conflicting design objectives; and (iii) the controller design is translated into a convex optimization problem, guaranteeing the efficient computation of its global optimum. Apart from extensive numerical evaluation, the potential of the design methodology is experimentally illustrated on an active air bearing setup.

This monograph is the result of four years of PhD research at the Division of Production Engineering, Machine Design & Automation (PMA), Department of Mechanical Engineering, Katholieke Universiteit Leuven, Belgium. I am grateful to all people who contributed to this work. I would like to give a special word of thanks to my supervisors Jan Swevers and Bram Demeulenaere for offering me support and guidance, while at the same time giving me the freedom to choose my research niche. I wish to acknowledge Z. Liu and Prof. L. Vandenberghe (UCLA, Electrical Engineering Department) for their kind assistance and pertinent comments concerning the numerical solution of the SDPs involved. Also many thanks to the Research Foundation–Flanders (FWO–Vlaanderen) for providing a fellowship for this research. Last but not least, I am indebted to my parents who always encouraged and supported me.

Leuven, Belgium,   Goele Pipeleers
August 2009

Contents

Abbreviations

Sets

$\mathbf{R}$	real numbers
$\mathbf{R}_n$	real n-vectors
$\mathbf{R}_{n \times m}$	real $n \times m$ matrices
$\mathbf{C}$	complex numbers
$\mathbf{C}_n$	complex n-vectors
$\mathbf{C}_{n \times m}$	complex $n \times m$ matrices
$\mathbf{S}_n$	symmetric $n \times n$ matrices
$\mathbf{H}_n$	Hermitian $n \times n$ matrices

General Symbols

I_n	$n \times n$ identity matrix, where the subscript is omitted if the dimension is clear from the context
$0_{n \times m}$	$n \times m$ zero matrix, where the subscript is omitted if the dimension is clear from the context
X^T	transpose of matrix X
X^H	Hermitian (complex conjugate) transpose of matrix X
$\mathrm{Tr}\{X\}$	trace of matrix X
$\otimes$	matrix Kronecker product
$|X|$	cardinal number of finite set X
$\sigma_{\max}\{X\}$	largest singular value of matrix X
$\Re\{X\}$	real part of X
$\Im\{X\}$	imaginary part of X

Discrete-time, Linear Time-invariant Systems

f_s sample frequency [Hz]
T_s sample period [s]: $T_s = 1/f_s$
q one-sample-advance operator
z discrete-time Laplace variable
k index labeling the sampled time instants kT_s
ω frequency [rad/s]
$X(q)$ difference equation of system X
$X(z)$ transfer function (matrix) of system X
$X(\omega)$ FRF (matrix) of system X
$X_+(z)$ noninvertible part of SISO transfer function $X(z)$, comprising a delay
                     equal to the relative degree of $X(z)$ and its nonminimum-phase zeros
$X_-(z)$ invertible part of SISO transfer function $X(z)$: $X(z) = X_+(z)X_-(z)$
$\|X(z)\|_\infty$ $\mathcal{H}_\infty$ norm of system X

Control Configuration

$w(k)$ exogenous input
$u(k)$ control input
$v(k)$ regulated output
$y(k)$ measured output
$r(k)$ reference input
$d(k)$ disturbance input
$\eta(k)$ plant output
$e(k)$ tracking error: $e(k) = r(k) - \eta(k)$
G plant
P generalized plant
K controller
H closed-loop system
S closed-loop sensitivity
T closed-loop complementary sensitivity

Periodic Input

T_p nominal value of the period [s]
f_p nominal value of the fundamental frequency [Hz]: $f_p = 1/T_p$
ω_p nominal value of the fundamental frequency [rad/s]: $\omega_p = 2\pi f_p$
$\boldsymbol{\delta}$ relative uncertainty on f_p and ω_p
$f_{p,\delta}$ potential values of the fundamental frequency [Hz]: $f_{p,\delta} = f_p(1 + \delta)$,
                     where $|\delta| \leq \boldsymbol{\delta}$
$\omega_{p,\delta}$ potential values of the fundamental frequency [rad/s]: $\omega_{p,\delta} = 2\pi f_{p,\delta}$
l index labeling the harmonics
$\mathcal{L}$ set of harmonics to be suppressed

$n_{\mathscr{L}}$	number of elements in $\mathscr{L}$
W_l	positive weight quantifying the relative importance of the l'th harmonic in the periodic input
Ω_l	uncertainty interval on the l'th harmonic frequency [rad/s]: $\Omega_l = \left[l\omega_\mathrm{p}(1-\delta)\,,\ l\omega_\mathrm{p}(1+\delta) \right)$
Λ	signal generator of the periodic input, for nominal period T_p
n_Λ	order of Λ

Abbreviations

FIR	finite impulse response
FRF	frequency response function
KYP	Kalman-Yakubovich-Popov
LMI	linear matrix inequality
LTI	linear time-invariant
MIMO	multiple-input multiple-output
rms	root-mean-square
SDP	semi-definite programming problem
SISO	single-input single-output
SOCP	second-order cone problem

Chapter 1
Introduction

1.1 Motivation

1.1.1 Periodic Inputs Deserve Special Attention

Periodic reference and disturbance signals are widespread in engineering practice, as every rotating machine and repeated process involves periodicity. Periodic disturbances are for instance encountered in the track-following servo system of disk drives [24, 25, 26, 101], steel casting [98], power electronics [12, 164, 165], active air bearing systems [7, 67], active noise control [11], satellite attitude stabilization [16, 156], peristaltic pumps used in medical devices [64] and robotized laparoscopic surgery [50]. Furthermore, disturbances due to rotating unbalances or nearby combustion engines are dominantly periodic. On the other hand, periodic reference trajectories occur in noncircular machining [83, 148], electronic cam motion generation [84] and robots performing repetitive tasks [80, 119].

In engineering practice, better performing controllers are an essential complement to improved machine design in the continual quest for better tracking and disturbance rejection performance. The attainable performance of a controller is, however, bounded by measurement noise, model inaccuracies, actuator saturation, etc., and in face of these limitations, exploiting all knowledge available on the reference and disturbance inputs is indispensable to achieve the tightening performance demands.

1.1.2 Linear Control for Periodic Inputs

In linear control theory, the scope of this monograph, considerable effort has been devoted to specialized controller designs for periodic inputs. This control problem is often handled in the context of *output regulation*, which concerns the design of an internally stabilizing controller that yields perfect asymptotic rejection/tracking of

G. Pipeleers et al.: Optimal Linear Controller Design for Periodic Inputs, LNCIS 394, pp. 1–4.
springerlink.com   © Springer-Verlag Berlin Heidelberg 2009

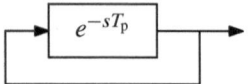

Fig. 1.1 Universal generator of signals with period T_p.

persistent input signals. Such signals comprise infinite energy signals of which the excitation frequencies are known, and mathematically, they are most conveniently described as the autonomous output of a marginally stable system. This system is commonly referred to as exosystem or signal generator, where the latter terminology is adopted here. Figure 1.1 shows a universal periodic signal generator, which, determined by its initial conditions, can generate any signal with period T_p [s].

In the early 1970s, Davison *et al.* [34, 35] and Francis *et al.* [45, 46, 47, 48] laid the foundation of regulation theory with the Internal Model Principle. This principle states that perfect asymptotic rejection/tracking of persistent inputs can only be attained by replicating the signal generator in a stable feedback loop. In this earliest form, the Internal Model Principle considers the classical feedback control configuration where both the regulated output and controller input correspond to the tracking error.

In the 1980s, Inoue *et al.* [75, 76], Hara *et al.* [59, 60] and Tomizuka *et al.* [145, 146] translated the Internal Model Principle into a feedback controller design that achieves output regulation of periodic inputs of which only the period T_p is known. The controllers are called *repetitive controllers*, and their structure explicitly incorporates the signal generator shown in Figure 1.1. Ever since its origin, repetitive control received continual interest in the literature, where contributions involve both theoretical improvements and practical applications (see e.g. [33, 66] for a survey).

Although specializing a *feedforward controller* design for periodic inputs is beneficial for nonminimum-phase or uncertain systems, it is only sparsely covered in the literature [143, 152]. Feedforward control is usually applied to a reference input for improving the overall tracking performance, but by combining it with a disturbance observer (see e.g. [103]), feedforward control can also be applied to improve disturbance attenuation. The resulting control strategy is, however, feedback in nature and therefore it is here referred to as *estimated disturbance feedback control*. Similar to feedforward control, only few contributions deal with exploiting the periodic input characteristics in an estimated disturbance feedback controller design [144, 151].

In the 1990s, fundamental research on the output regulation problem regained interest. Among others, contributions involve relaxing the assumption in the Internal Model Principle that the regulated output constitutes the controller input, and combining regulation with additional performance specifications and input constraints [117]. In addition, *feedback controllers* for periodic inputs are proposed that not only exploit the input periodicity, as repetitive controllers do, but also exploit the input's harmonic frequency content (i.e., the frequencies of the harmonic components present in the input) [65, 88]. According to the Internal Model Principle, such feedback controllers include a system that can generate periodic signals with a *given* harmonic frequency content. Repetitive controllers, on the other hand, include the universal periodic signal generator of Figure 1.1, which can generate periodic signals with *any* harmonic frequency content, and hence, constitute a subclass of feedback controllers.

To conclude, the literature on linear controller design for periodic inputs distinguishes four approaches: feedforward control, estimated disturbance feedback control, repetitive control and feedback control, and has gained a marked status in modern control literature.

1.1.3 Problem Statement

Although the literature on controller design for periodic inputs is vast, some issues remain open:

- The majority of the results, for all aforementioned control strategies, rely on the assumption that the period of the input is accurately known or measurable. In practice, however, this assumption is often jeopardized by clock error drift, jitter, measurement noise, disturbances in the period control loop, etc.
- Many controller designs adopt a single-objective point of view to the considered control problem with main (sole) emphasis on rejecting/tracking the periodic inputs. However, this improved periodic performance often compromises other performance aspects of the closed-loop system, such as the attenuation of nonperiodic disturbances, transient response time, etc., invoking the need for a multi-objective design philosophy.
- Except between feedforward and estimated disturbance feedback control, the literature reveals little interplay between the different control strategies. For instance, the intuitive feedforward controller design proposed in [152] is not investigated within the recent theoretical findings presented in [117], and while many applications mark the practical relevance of repetitive control, the performance of such controllers is not compared to feedback controllers that exploit the input's harmonic frequency content.

1.2 Contribution

This monograph presents a *general design methodology* for controllers facing periodic inputs. The methodology relies on the following keystones and hereby contrasts the existing design approaches:

Periodic Performance Index: Closed-loop periodic performance is quantified in terms of a periodic performance index, which explicitly accounts for period-time uncertainty. Hence, low values of this performance index translate into good closed-loop attenuation/tracking of the periodic inputs even if the actual period differs from its nominal value.

Multi-objective Control: Although improving the closed-loop periodic performance is the main focus of the controller design, the design methodology is multi-objective in nature and allows incorporating a variety of additional design specifications.

Convex Optimization: To guarantee the reliable and efficient computation of its global optimum, the multi-objective controller design problem is translated into a convex optimization problem. This transformation is enabled by the Youla parametrization and besides the efficient computation of the global optimum, the convexity of the obtained optimization problem facilitates the generation of trade-off curves between conflicting performance specifications. These curves indicate fundamental limits of performance in the controller design (performance bounds no controller can break), and will be shown a valuable design tool.

While this design methodology in itself is yet innovative, it has the additional contribution of bridging the current gap between the different control strategies. The methodology can be translated into a feedforward controller, estimated disturbance feedback controller, repetitive controller and feedback controller design, and hereby it relates these control strategies to a common ground, emphasizing their mutual relations.

Apart from extensive numerical evaluation, the potential of the design methodology is experimentally illustrated on an active air bearing setup.

1.3 Outline

Chapter 2 presents the developed controller design methodology and details its fundamental concepts. The four subsequent chapters each apply the design methodology to a specific control strategy: Chapter 3 starts with feedforward control, while Chapter 4 handles the estimated disturbance feedback controller design with the developed design methodology. Chapter 5 deals with the repetitive controller design, after which Chapter 6 applies the methodology to design feedback controllers that exploit the input's harmonic frequency content. Each of these chapters is provided with numerical results, which emphasize the capability of the presented design methodology to reproduce and outperform major current design approaches, for each of the four control strategies. Chapter 7 presents the experimental validation of the methodology, where it is applied to design a repetitive controller for an active air bearing setup. Chapter 8 summarizes the conclusions of this monograph.

Chapter 2
Design Methodology for Controllers Facing Periodic Inputs

2.1 Introduction

This chapter presents the general design methodology for controllers facing periodic input signals and elaborates on its fundamental concepts. The subsequent chapters apply this methodology to specific control strategies from the literature and hereby yield a more concrete impression of its functionalities and potential.

The design methodology starts from the general control configuration, see e.g. [13, 131, 163], which constitutes a universal way of formulating control problems. Section 2.2 details this control configuration together with the required *a priori* knowledge of the periodic input. Fundamental to the methodology is its multi-objective nature, combining improved closed-loop periodic performance with additional design specifications. Section 2.3 elaborates on the formulation of multi-objective control problems and defines the periodic performance index, the mathematical means to incorporate good periodic performance in the controller design. The efficient and reliable solution of the resulting multi-objective design problem relies on the benefits of convex optimization (Section 2.5), where the convex reformulation of the controller design problem is enabled by the Youla parametrization (Section 2.4).

The presented methodology is restricted to the design of a linear time-invariant (LTI) controller for an LTI system. Complying with the literature on control for periodic inputs, this monograph emphasizes the single-input single-output (SISO) discrete-time controller design. This restriction is however not fundamental to the methodology and Section 2.6 discusses the extension to continuous-time and multiple-input multiple-output (MIMO) control.

Concerning notation on discrete-time systems: the sample frequency is denoted by f_s [Hz], where $T_s = 1/f_s$ [s] indicates the sample period and index k labels the sampled time instants kT_s. The one-sample-advance operator is indicated by q, while z denotes the discrete-time Laplace variable. For a discrete-time LTI system P, $P(q)$ and $P(z)$ correspond to its difference equation and transfer function matrix, respectively, while its frequency response function (FRF) matrix is denoted by $P(\omega)$

G. Pipeleers et al.: Optimal Linear Controller Design for Periodic Inputs, LNCIS 394, pp. 5–21.
springerlink.com

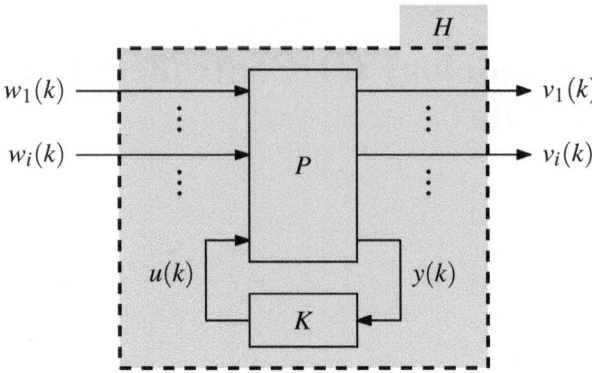

Fig. 2.1 General control configuration, where controller K is designed for generalized plant P, yielding closed-loop system H. Signals $u(k)$ and $y(k)$ respectively correspond to the plant's control input and measured output, while exogenous inputs $w_i(k)$ and regulated outputs $v_i(k)$ are used to define multiple design specifications.

instead of $P\left(e^{j\omega T_s}\right)$ to alleviate notation. To differentiate between variables, in this monograph, a system P is commonly indicated by $P(q)$ or $P(z)$, where the abuse of notation involves omitting the initial conditions.

2.2 Control Problem Formulation

The control problem is formulated in terms of the general control configuration discussed in Section 2.2.1, while Section 2.2.2 details the *a priori* knowledge of the periodic input required by the controller design.

2.2.1 General Control Configuration

Figure 2.1 shows the general control configuration, which provides a universal way of formulating control problems. The generalized plant $P(z)$ constitutes a mathematical model of the system to be controlled. Its input signals manipulable by the controller are grouped in the control input $u(k)$, while the measured output $y(k)$ comprises the plant outputs accessible to the controller. The controller is denoted $K(z)$ and its design must guarantee internal stability of the closed-loop system $H(z)$ and make it behave in a desired manner. Desired closed-loop behavior is translated into multiple design specifications, labeled by index i, where the i'th specification involves the closed-loop subsystem $H_i(z)$ from exogenous input $w_i(k)$ to regulated output $v_i(k)$. Exogenous inputs generally correspond to a disturbance, noise input or reference command, while regulated outputs are usually signals that should be rendered "small", such as the tracking error or control effort. Subsystems $H_i(z)$ correspond to the (block-) diagonal components of $H(z)$:

$$\begin{bmatrix} v_1(k) \\ \vdots \\ v_i(k) \\ \vdots \end{bmatrix} = \underbrace{\begin{bmatrix} H_1(q) & \cdots & \star & \cdots \\ \vdots & \ddots & \vdots & \\ \star & \cdots & H_i(q) & \\ \vdots & & & \ddots \end{bmatrix}}_{H(q)} \begin{bmatrix} w_1(k) \\ \vdots \\ w_i(k) \\ \vdots \end{bmatrix} ,$$

and specifying closed-loop performance in terms of $H_i(z)$ solely, disregarding the off-diagonal subsystems indicated by a $\star$, may require the recurrence of exogenous inputs and regulated outputs. Since multiple design objectives are considered, it is advisable to scale the exogenous inputs and regulated outputs to their maximum expected or allowed value [131]. In a SISO control problem, signals $u(k)$ and $y(k)$ are scalar, while vector-valued exogenous inputs and regulated outputs are still allowed.

Provided that the generalized plant is decomposed as follows:

$$\begin{bmatrix} v_i(k) \\ y(k) \end{bmatrix} = \begin{bmatrix} P_i(q) & P_{iu}(q) \\ P_{yi}(q) & P_{yu}(q) \end{bmatrix} \begin{bmatrix} w_i(k) \\ u(k) \end{bmatrix} ,$$

closed-loop subsystem $H_i(z)$ is given by

$$H_i(z) = P_i(z) + P_{iu}(z) K(z) \left[I - P_{yu}(z) K(z) \right]^{-1} P_{yi}(z) . \tag{2.1}$$

If the plant model $P(z)$ is uncertain, a robust controller design is demanded. To that end, an uncertainty set $\mathbf{\Delta}$ is specified that captures the uncertainty Δ on the nominal plant model $P(z)$. A robust controller not only performs well for the nominal model $P(z)$ but for all potential plant models, indicated by $P_\Delta(z)$, where $\Delta \in \mathbf{\Delta}$. Following the same notation, a given controller $K(z)$ gives rise to a set of potential closed-loop systems $H_\Delta(z)$, while $H(z)$ indicates the nominal closed-loop system, corresponding to the nominal plant $P(z)$.

While alternative uncertainty sets are allowed, this monograph focusses on multiplicative unstructured plant uncertainty [131]. Hereby, the generalized plant $P_\Delta(z)$ depends on an uncertain system $G_\Delta(z)$ of the form

$$G_\Delta(z) = G(z) \left[1 + W_G(z) \Delta(z) \right] , \quad \Delta(z) \in \mathbf{\Delta} . \tag{2.2a}$$

Transfer function $G(z)$ corresponds to the nominal model, while uncertainty set $\mathbf{\Delta}$ is given by

$$\mathbf{\Delta} = \{ \Delta(z) \text{ is a stable system with } \|\Delta(z)\|_\infty \leq 1 \} , \tag{2.2b}$$

and stable transfer function $W_G(z)$ determines the "size" of the uncertainty.

2.2.2 Periodic Input

Inspired by the literature, this monograph focusses on a single, scalar periodic input signal, while the relaxation of this restriction is discussed in Section 2.6. Good closed-loop rejection/tracking of the periodic input corresponds to the i_p'th design specification $(i = i_p)$, where the subscript $(\cdot)_{i_p}$ is shortened to $(\cdot)_p$. Hence, $w_p(k)$ constitutes the periodic input and its effect on $v_p(k)$, in closed loop related to $w_p(k)$ by subsystem $H_p(z)$, should be reduced/eliminated.

The nominal value of the input period is denoted by T_p [s], where $f_p = 1/T_p$ [Hz] indicates the corresponding fundamental frequency and $\omega_p = 2\pi f_p$ [rad/s]. Index l labels the harmonics of the periodic input, where $0 \le l \le T_p f_s/2$ is assumed without loss of generality. The set of harmonics l that are present in $w_p(k)$ and should be suppressed by the controller is denoted by $\mathscr{L}$, and $n_{\mathscr{L}}$ equals the number of elements in $\mathscr{L}$. To each harmonic $l \in \mathscr{L}$, a positive weight W_l is attributed, quantifying its relative importance in $w_p(k)$. Hence, W_l generally corresponds to the amplitude of the corresponding Fourier coefficient, or a (rough) estimate thereof.

The design methodology allows accounting for period-time uncertainty, which is modeled as relative (multiplicative) uncertainty on ω_p, bounded by $\boldsymbol{\delta}$. Hence, all potential values $\omega_{p,\delta}$ of the fundamental frequency are given by

$$\omega_{p,\delta} = \omega_p(1+\delta), \quad |\delta| \le \boldsymbol{\delta}, \tag{2.3}$$

while

$$\Omega_l = \left[l\omega_p(1-\boldsymbol{\delta}), \, l\omega_p(1+\boldsymbol{\delta})\right] \tag{2.4}$$

equals the corresponding uncertainty interval on the l'th harmonic frequency. According to $\omega_{p,\delta}$: $f_{p,\delta} = \omega_{p,\delta}/(2\pi)$ and $T_{p,\delta} = 1/f_{p,\delta}$.

Implied by the Internal Model Principle [34, 45, 46, 47, 48], a controller that yields perfect asymptotic rejection of $w_p(k)$ must include the corresponding signal generator. The generator of $w_p(k)$ for nominal period T_p is denoted by $\Lambda(z)$ and comprises the signal generators of each of its harmonic components:

$$\Lambda(z) = \prod_{l \in \mathscr{L}} \Lambda_l(z), \tag{2.5a}$$

where

$$\Lambda_l(z) = \frac{z^2}{\left(z - e^{-jl\omega_p T_s}\right)\left(z - e^{jl\omega_p T_s}\right)}, \quad \forall l \in \mathscr{L} \setminus \{0, T_p f_s/2\}, \tag{2.5b}$$

$$\Lambda_0(z) = \frac{z}{(z-1)}, \tag{2.5c}$$

$$\Lambda_{T_p f_s/2}(z) = \frac{z}{(z+1)}. \tag{2.5d}$$

The order of $\Lambda(z)$ is indicated by n_Λ and equals

$$n_\Lambda = 2n_{\mathscr{L}} - \left| \mathscr{L} \cap \{0, T_p f_s/2\} \right| , \qquad (2.6)$$

where for a finite set X, its cardinal number $|X|$ equals the number of elements in X. If T_p contains an integer number N of sample periods and all harmonics are present in $w_p(k)$:

$$\Lambda(z) = \frac{z^N}{z^N - 1} ,$$

and $n_\Lambda = N$.

2.3 Optimal Controller Design

The objective is to design an internally stabilizing controller $K(z)$ that optimizes the closed-loop periodic performance, i.e., minimizes the steady-state effect of $w_p(k)$ on $v_p(k)$, but without being blind for the controller's effect on conflicting performance aspects. As this design philosophy translates into a multi-objective controller design, Section 2.3.1 first discusses the formal structure of such a design problem. The periodic performance index, defined in Section 2.3.2, quantifies closed-loop periodic performance and allows incorporating good suppression/tracking of the periodic input in the controller design. To conclude, Section 2.3.3 presents a multi-objective control problem that often recurs in this monograph.

2.3.1 Formal Multi-objective Control Problem

The methodology allows incorporating any design specification provided that it is closed-loop convex, that is: the set of LTI systems satisfying the specification is convex. As detailed in e.g. [13, 14, 122], most performance and robustness specifications in linear control are closed-loop convex, including internal stability, system norms, convex pole placement constraints, constraints on a closed-loop step response such as asymptotic tracking, under- and overshoot, settling and rise-time, etc.

Design specifications either impose a hard constraint on the controller design or involve a performance aspect that should be optimized. For instance, internal closed-loop stability is an indispensable hard constraint, where the convex set of internally stable closed-loop systems is denoted by $\mathscr{S}_{stab}$. Additional hard design constraints generally relate to fixed performance targets or robust closed-loop stability, as imposed by the small gain theorem or translated into stability margins like gain and phase margin. Indices i related to these additional design constraints are grouped into the set $\mathscr{I}_{constr}$, while each constraint can formally be described as $H_i(z) \in \mathscr{S}_i$, where the set $\mathscr{S}_i$ is convex.

In contrast to hard constraints, so-called soft design constraints involve a performance specification that should be optimized but for which no *a priori* target is given. Mathematically, such a performance specification is translated into a performance index γ_i that should be rendered small. Indices i related to such soft design constraints are grouped into the set $\mathscr{I}_{\mathrm{perf}}$, and formally, γ_i is defined by $\big(H_i(z), \gamma_i\big) \in \mathscr{S}_i$, where set $\mathscr{S}_i$ is convex. To minimize all performance indices involved, they are combined into a weighted sum:

$$f_{\mathrm{obj}} = \sum_{i \in \mathscr{I}_{\mathrm{perf}}} \alpha_i \gamma_i \,,$$

where weights $\alpha_i \geq 0$ reflect the relative "importance" attributed to the various performance specifications. This approach is referred to as scalarization [15] and translates into the following formulation of the multi-objective controller design problem:

$$
\begin{aligned}
\underset{K(z),\gamma_i}{\text{minimize}} \quad & \sum_{i \in \mathscr{I}_{\mathrm{perf}}} \alpha_i \gamma_i \\
\text{subject to} \quad & \big(H_i(z), \gamma_i\big) \in \mathscr{S}_i \,, && \forall i \in \mathscr{I}_{\mathrm{perf}} \\
& H_i(z) \in \mathscr{S}_i \,, && \forall i \in \mathscr{I}_{\mathrm{constr}} \\
& H(z) \in \mathscr{S}_{\mathrm{stab}} \,.
\end{aligned}
$$

2.3.2   Periodic Performance Index

Closed-loop periodic performance relates to the steady-state effect of $w_{\mathrm{p}}(k)$ on $v_{\mathrm{p}}(k)$, and to incorporate good closed-loop periodic performance in the controller design, it is quantified by the periodic performance index γ_{p}, where the shortened notation $(\cdot)_{\mathrm{p}}$ instead of $(\cdot)_{i_{\mathrm{p}}}$, of Section 2.2.2 is continued.

To quantify closed-loop periodic performance, in a first step the FRF of $H_{\mathrm{p}}(z)$ is analyzed around each harmonic $l \in \mathscr{L}$ and V_l is determined as follows:

$$V_l = W_l \max_{\omega \in \Omega_l} \big\{ |H_{\mathrm{p}}(\omega)| \big\} \,.$$

If W_l equals the amplitude of the l'th Fourier coefficient of $w_{\mathrm{p}}(k)$, V_l corresponds to the worst-case amplitude of the l'th Fourier coefficient of $v_{\mathrm{p}}(k)$ over all potential $\omega_{\mathrm{p},\delta}$ values (2.3). More generally, V_l corresponds to the worst-case steady-state closed-loop reduction of the l'th harmonic from $w_{\mathrm{p}}(k)$ to $v_{\mathrm{p}}(k)$, weighted by W_l.

In a second step, the V_l values for all $l \in \mathscr{L}$ are combined into a scalar performance index. Performing this combination in an ∞-norm based manner yields

$$\gamma_{\mathrm{p}} = \max_{l \in \mathscr{L}} \{V_l\} \,, \tag{2.7a}$$

$$= \max_{l \in \mathscr{L}} \Big\{ W_l \max_{\omega \in \Omega_l} \big\{ |H_{\mathrm{p}}(\omega)| \big\} \Big\} \,, \tag{2.7b}$$

while an alternative definition is based on the 2-norm[1]:

$$\gamma_{p,2} = \sqrt{\sum_{l \in \mathscr{L}} V_l^2} \,, \tag{2.8a}$$

$$= \sqrt{\sum_{l \in \mathscr{L}} \left[W_l \max_{\omega \in \Omega_l} \{ |H_p(\omega)| \} \right]^2} \,. \tag{2.8b}$$

Since in the subsequent chapters, definition (2.7) is more often used, the corresponding index is denoted by γ_p instead of $\gamma_{p,\infty}$. For both definitions, lower index values indicate better closed-loop asymptotic attenuation/tracking of the periodic input $w_p(k)$ for all $\omega_{p,\delta}$. Perfect asymptotic attenuation/tracking of $w_p(k)$ is referred to as perfect periodic performance, and yields $\gamma_p = \gamma_{p,2} = 0$. Although definitions (2.7) and (2.8) are closely related by the equivalence of vector norms, see e.g. [131]:

$$\gamma_p \leq \gamma_{p,2} \leq \sqrt{n_{\mathscr{L}}} \, \gamma_p \,,$$

their interpretations are subtly different. A given γ_p value guarantees worst-case closed-loop attenuation of each harmonic l by a level γ_p/W_l, where the weights W_l can be regarded as tuning parameters to distribute control effort over the harmonics. On the other hand, $\gamma_{p,2}$ is more appropriate if the input spectrum is known, since it bounds from above the worst-case root-mean-square (rms) value of $v_p(k)$ over all $\omega_{p,\delta}$:

$$\max_{|\delta| < \delta} \underbrace{\sqrt{\sum_{l \in \mathscr{L}} \left(W_l |H_p(l\omega_{p,\delta})| \right)^2}}_{\mathrm{rms}(v_p(k)) \text{ for } \omega_{p,\delta}} \leq \gamma_{p,2} \,, \tag{2.9}$$

provided that W_l are set equal to the amplitudes of the corresponding Fourier coefficients[2].

In view of the formal multi-objective control problem stated in the previous section, good closed-loop periodic performance can be added as either a hard or a soft constraint. In the former case γ_p is fixed, whereas in the latter case γ_p constitutes an optimization variable. Definition (2.7) of γ_p corresponds to the following set $\mathscr{S}_p$:

$$\mathscr{S}_p = \left\{ (H_p(z), \gamma_p) \,\middle|\, W_l |H_p(\omega)| \leq \gamma_p \,, \forall \omega \in \Omega_l \text{ and } \forall l \in \mathscr{L} \right\} \,,$$

whereas for definition (2.8):

---

[1] While only the ∞-norm and 2-norm based definitions are used in this monograph, any vector norm can be used to combine the V_l values into a scalar performance index, yielding a closed-loop convex periodic performance specification.

[2] The left-hand side of (2.9) is also closed-loop convex. However, converting it into a set of LMIs by the generalized KYP, similar to the procedure of Section 2.5, yields an SDP much harder to solve compared to definition (2.8).

$$\mathscr{S}_{\mathrm{p},2} = \left\{ \left(H_{\mathrm{p}}(z), \gamma_{\mathrm{p},2}\right) \middle| \sqrt{\sum_{l \in \mathscr{L}} V_l^2} \le \gamma_{\mathrm{p},2} \text{ , where } \forall l \in \mathscr{L} : \right.$$

$$\left. W_l |H_{\mathrm{p}}(\omega)| \le V_l \, , \, \forall \omega \in \Omega_l \right\} .$$

2.3.3 Control Problem of Particular Interest

One particular instance of the multi-objective control problem often recurs in the subsequent chapters. In this problem, closed-loop periodic performance is traded-off against closed-loop nonperiodic performance, i.e., the closed-loop attenuation/tracking of nonperiodic inputs. Good nonperiodic performance corresponds to the i_{np}'th design specification and notation $(\cdot)_{i_{\mathrm{np}}}$ is shortened to $(\cdot)_{\mathrm{np}}$. While alternative definitions for the nonperiodic performance index γ_{np} are allowed, the following one is used in this monograph:

$$\gamma_{\mathrm{np}} = \|H_{\mathrm{np}}(z)\|_{\infty} \, , \tag{2.10}$$

yielding

$$\mathscr{S}_{\mathrm{np}} = \left\{ \left(H_{\mathrm{np}}(z), \gamma_{\mathrm{np}}\right) \middle| |H_{\mathrm{np}}(\omega)| \le \gamma_{\mathrm{np}} \, , \, \forall \omega \in [0, \pi f_{\mathrm{s}}] \right\} .$$

To account for the amplitude spectrum of nonperiodic input $w_{\mathrm{np}}(k)$, an appropriate weighting function should be included in $H_{\mathrm{np}}(z)$ [131].

To investigate the trade-off between γ_{p} and γ_{np}, the following design problem is solved for various weights $\alpha \ge 0$:

$$\begin{aligned}
\underset{K(z), \gamma_{\mathrm{p}}, \gamma_{\mathrm{np}}}{\text{minimize}} \quad & \gamma_{\mathrm{p}} + \alpha \gamma_{\mathrm{np}} \\
\text{subject to} \quad & \left(H_{\mathrm{p}}(z), \gamma_{\mathrm{p}}\right) \in \mathscr{S}_{\mathrm{p}} \\
& \left(H_{\mathrm{np}}(z), \gamma_{\mathrm{np}}\right) \in \mathscr{S}_{\mathrm{np}} \\
& H(z) \in \mathscr{S}_{\mathrm{stab}} \, .
\end{aligned}$$

In many applications, performance indices γ_{p} and γ_{np} are conflicting, implying that improved periodic performance (lower γ_{p} value) comes at the price of degraded nonperiodic performance (higher γ_{np} value). This conflicting behavior stems from the Bode Integral Theorem [10, 21, 22, 49, 69, 138], and prevails in applications where $H_{\mathrm{p}}(z) = H_{\mathrm{np}}(z)$. For a stable, SISO discrete-time system $R(z)$ with gain ρ, minimum-phase zeros $z_{-,i}$, nonminimum-phase zeros $z_{+,i}$ and stable poles p_i:

$$R(z) = \rho \frac{\prod_{i=1}^{m_-}(z - z_{-,i}) \prod_{i=1}^{m_+}(z - z_{+,i})}{\prod_{i=1}^{n}(z - p_i)} \, ,$$

the Bode Integral Theorem states that

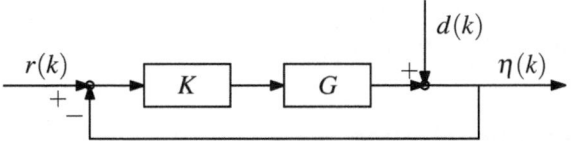

Fig. 2.2 Classical feedback control configuration, where feedback controller K should make output $\eta(k)$ of plant G follow reference trajectory $r(k)$ in the presence of output disturbance $d(k)$.

$$\frac{1}{\pi f_s}\int_0^{\pi f_s}\ln|R(\omega)|d\omega = \ln|\rho| + \sum_{i=1}^{m_+}\ln|z_{+,i}| . \tag{2.11}$$

In many control applications, fundamental limitations dictate unity gain ($\rho = 1$) in the closed-loop transfer function, independent of the controller design. As this bounds the right-hand side of (2.11) to be larger than or equal to zero, a trade-off between γ_p and γ_{np} emerges: pushing the closed-loop FRF down to zero at the periodic input's harmonics yields a negative contribution in the integral (2.11) and hereby invokes a positive contribution at intermediate frequencies, which corresponds to closed-loop amplification.

The most well-known control application that features the fundamental unity gain limitation ($\rho = 1$) is the classical feedback control system, shown in Figure 2.2. $K(z)$ is the controller to be designed and its purpose is to make output $\eta(k)$ of plant $G(z)$ follow the reference trajectory $r(k)$ in the presence of output disturbance $d(k)$. Performance is determined by the closed-loop sensitivity $S(z)$, as this transfer function determines the tracking error $e(k) = r(k) - \eta(k)$:

$$S(q) = \frac{e(k)}{r(k) - d(k)} = \frac{1}{1 + K(q)G(q)} .$$

If the plant is strictly causal (denominator order larger than numerator order), any causal controller yields a sensitivity function with unity gain. Hence, if $r(k) - d(k)$ comprises both a periodic and nonperiodic signal (that is: $r(k) - d(k) = w_p(k) + w_{np}(k)$, while $v_p(k) = v_{np}(k) = e(k)$, yielding $H_p(z) = H_{np}(z) = S(z)$), γ_p and γ_{np} are conflicting.

Theoretically, the trade-off between γ_p and γ_{np} can be circumvented by creating very sharp notches at the harmonics, hereby making the negative contribution in (2.11) infinitely small. However, this comes at the expense of a very sluggish transient response[3] and relies on very accurate knowledge of the input period (see Sections 5.4.2 and 6.5.2). While the latter assumption is often jeopardized in practice by measurement noise, clock error drift, jitter, etc., sluggish transient responses are generally unacceptable. Taking into account these practical issues, in many control

---

[3] This holds for an LTI controller design, as is considered in this monograph, but can be circumvented by using time-varying control [87].

applications the Bode Integral Theorem dictates a trade-off between closed-loop periodic and nonperiodic performance.

2.4 Youla Parametrization

In general, multi-objective control problems are hard to solve, since despite the closed-loop convexity of the design specifications, they are nonconvex when formulated in terms of the controller parameters. The origin of this nonconvexity is twofold: (i) the closed-loop transfer function matrix $H(z)$ depends in a nonlinear way on the controller $K(z)$, see Equation 2.1; and (ii) parameterizing $K(z)$ as a transfer function or state-space model yields a nonlinear relation between the design parameters and the evaluation of $K(z)$ for a given z.

Currently, two approaches exist to reformulate a multi-objective control problem as a convex optimization problem. The first approach relies on the Lyapunov shaping paradigm [36, 121], whereas the second strategy applies the Youla parametrization [37, 90, 159, 160]. Although the Lyapunov shaping paradigm allows incorporating (almost) asymptotic regulation constraints [88, 121], it cannot cope with uncertainty on the input period and the periodic performance indices defined in Section 2.3.2. While this is yet a decisive reason to adopt the Youla parametrization, the subsequent chapters additionally show that the Youla parametrization is able to reproduce many of the controller structures for periodic inputs reported in the literature.

In the Youla parametrization approach, the nonlinear relation between $H(z)$ and $K(z)$ is circumvented by a nonlinear, one-by-one transformation of $K(z)$ into the so-called Youla parameter $X(z)$, which relates affinely to $H(z)$ (Section 2.4.1). The second source of nonlinearity is handled by parameterizing $X(z)$ in an affine manner; that is: as an affine combination of given transfer functions (Section 2.4.2).

2.4.1 Parametrization of $H(z)$

The Youla parametrization (see e.g. [97, 163] for more details) states that all realizable, internally stable closed-loop systems $H(z)$ can be expressed as

$$H(z) = \Phi_1(z) + \Phi_2(z)X(z)\Phi_3(z) , \tag{2.12}$$

where $X(z)$ is a free, stable transfer function, called the Youla parameter, which relates in a one-to-one relationship to the corresponding controller $K(z)$. The transfer matrices $\Phi_1(z)$, $\Phi_2(z)$ and $\Phi_3(z)$ are determined by the generalized plant $P(z)$, augmented with an arbitrary stabilizing controller. As relation (2.12) holds for the entire transfer function matrix $H(z)$, similar relations apply to the diagonal subsystems $H_i(z)$:

$$H_i(z) = \Phi_{1,i}(z) + \Phi_{2,i}(z)X(z)\Phi_{3,i}(z) . \tag{2.13}$$

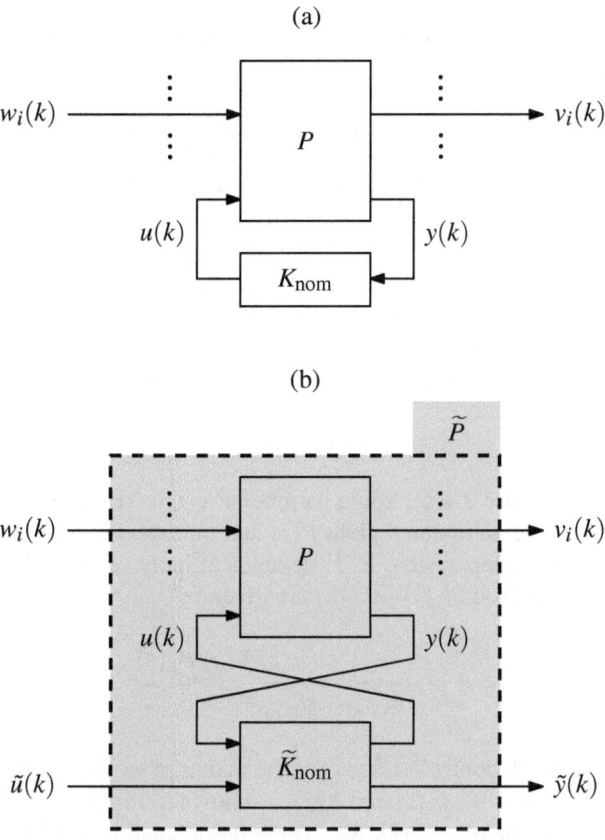

Fig. 2.3 (a) The construction of the Youla parametrization starts with the design of an arbitrary stabilizing feedback controller K_{nom}. (b) In a second step, K_{nom} is properly augmented to $\widetilde{K}_{\text{nom}}$ with auxiliary input $\tilde{u}(k)$ and auxiliary output $\tilde{y}(k)$, where the augmented plant $\widetilde{P}$ corresponds to the resulting closed-loop system with inputs $w_i(k)$, $\tilde{u}(k)$ and outputs $v_i(k)$, $\tilde{y}(k)$.

The derivation of $\Phi_1(z)$, $\Phi_2(z)$ and $\Phi_3(z)$ proceeds according to the steps illustrated in Figure 2.3. First, an arbitrary controller $K_{\text{nom}}(z)$, called the nominal controller, is designed, where the only requirement for $K_{\text{nom}}(z)$ is that it must yield an internally stable closed-loop system.

In a second step, $K_{\text{nom}}(z)$ is augmented to $\widetilde{K}_{\text{nom}}(z)$ so that it accepts an auxiliary input $\tilde{u}(k)$ and produces an auxiliary output $\tilde{y}(k)$. The resulting closed-loop system with inputs $w_i(k)$ and $\tilde{u}(k)$ and outputs $v_i(k)$ and $\tilde{y}(k)$ is called the augmented plant $\widetilde{P}(z)$, and its subsystems are indicated as follows:

$$\begin{bmatrix} v_i(k) \\ \tilde{y}(k) \end{bmatrix} = \begin{bmatrix} \widetilde{P}_i(q) & \widetilde{P}_{i\tilde{u}}(q) \\ \widetilde{P}_{\tilde{y}i}(q) & \widetilde{P}_{\tilde{y}\tilde{u}}(q) \end{bmatrix} \begin{bmatrix} w_i(k) \\ \tilde{u}(k) \end{bmatrix} .$$

The augmentation of $K_{\text{nom}}(z)$ must comply with three requirements: first, the transfer function from $y(k)$ to $u(k)$ in $\widetilde{K}_{\text{nom}}(z)$ must remain $K_{\text{nom}}(z)$:

$$\begin{bmatrix} u(k) \\ \tilde{y}(k) \end{bmatrix} = \underbrace{\begin{bmatrix} K_{\text{nom}}(q) & \widetilde{K}_{u\tilde{u}}(q) \\ \widetilde{K}_{\tilde{y}y}(q) & \widetilde{K}_{\tilde{y}\tilde{u}}(q) \end{bmatrix}}_{\widetilde{K}_{\text{nom}}(q)} \begin{bmatrix} y(k) \\ \tilde{u}(k) \end{bmatrix} .$$

Second, if $K_{\text{nom}}(z)$ is stable, the transfer functions $\widetilde{K}_{u\tilde{u}}(z)$ and $\widetilde{K}_{\tilde{y}y}(z)$ must be stable and invertible, whereas for unstable $K_{\text{nom}}(z)$, they must correspond to the product of a stable invertible transfer function and the unstable part of $K_{\text{nom}}(z)$ (see e.g. [97, 163]). Third, the design of $\widetilde{K}_{\tilde{y}\tilde{u}}(z)$ must guarantee that

$$\widetilde{P}_{\tilde{y}\tilde{u}}(z) = 0 . \tag{2.14}$$

As indicated in Figure 2.4(a), Youla parameter $X(z)$ is designed as a stable feedback controller for the augmented plant $\widetilde{P}(z)$ and on account of relation (2.14), the corresponding closed-loop system $H(z)$ depends affinely on $X(z)$. For instance, its diagonal components from $w_i(k)$ to $v_i(k)$ are given by

$$H_i(z) = \underbrace{\widetilde{P}_i(z)}_{\Phi_{1,i}(z)} + \underbrace{\widetilde{P}_{i\tilde{u}}(z)}_{\Phi_{2,i}(z)} X(z) \underbrace{\widetilde{P}_{\tilde{y}i}(z)}_{\Phi_{3,i}(z)} .$$

The resulting feedback controller $K(z)$ for the actual plant $P(z)$ corresponds to the feedback combination of $\widetilde{K}_{\text{nom}}(z)$ and $X(z)$, as shown in Figure 2.4(b).

In the case of a stable plant $P(z)$, $K_{\text{nom}}(z) = 0$ is a feasible nominal controller for which

$$\widetilde{K}_{\text{nom}}(z) = \begin{bmatrix} 0 & 1 \\ 1 & -P_{yu}(z) \end{bmatrix}$$

is a feasible augmentation. This yields

$$H_i(z) = \underbrace{P_i(z)}_{\Phi_{1,i}(z)} + \underbrace{P_{iu}(z)}_{\Phi_{2,i}(z)} X(z) \underbrace{P_{yi}(z)}_{\Phi_{3,i}(z)} ,$$

and comparison with Equation 2.1 reveals

$$X(z) = K(z)\left[I - P_{yu}(z)K(z)\right]^{-1} .$$

2.4.2 Parametrization of $X(z)$

Based on the results of the previous section, it is equivalent to design $X(z)$ instead of $K(z)$, which yields the benefit that $H(z)$ depends affinely on $X(z)$, (2.13), whereas its dependency on $K(z)$ is nonlinear (2.1). In order to obtain a convex

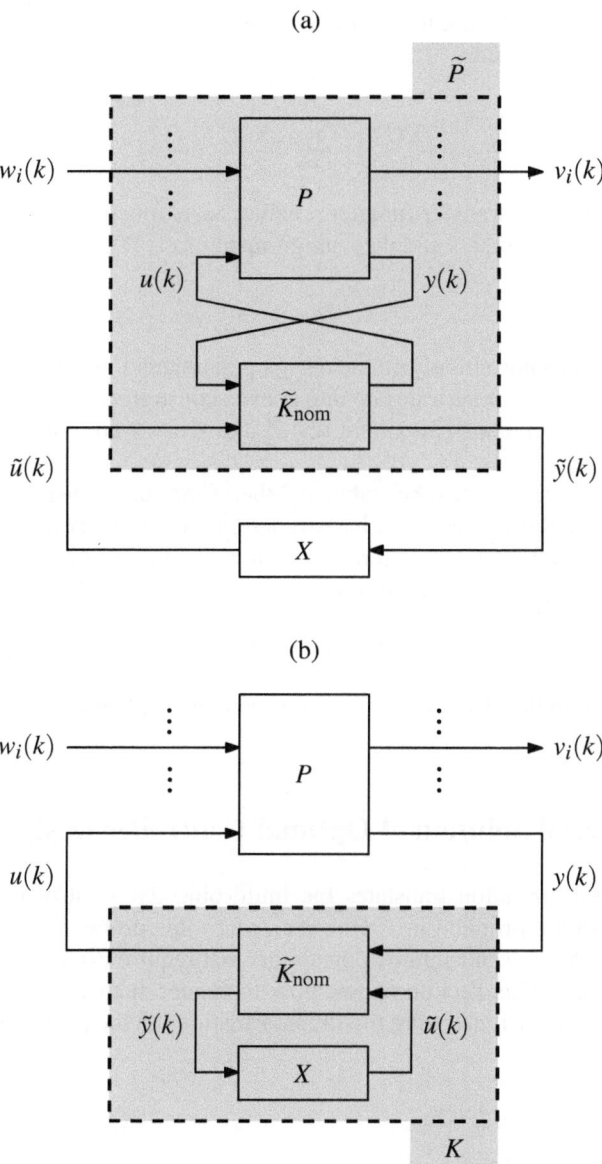

Fig. 2.4 (a) Youla parameter X is designed as a stable feedback controller for the augmented plant $\widetilde{P}$, where (b) the corresponding controller K for the actual plant P corresponds to the feedback combination of $\widetilde{K}_{\text{nom}}$ and X.

optimization problem, $X(z)$ must additionally be parameterized in an affine manner. That is: $X(z)$ is of the form

$$X(z) = \sum_{m=1}^{M} x_m X_m(z) \,, \tag{2.15}$$

where $X_m(z)$ are given transfer functions, called basis functions. Variables x_m constitute the design parameters and they are grouped in $x \in \mathbf{R}_M$:

$$x = \begin{bmatrix} x_1 & x_2 & \cdots & x_M \end{bmatrix}^T \,. \tag{2.16}$$

This way, $H(z)$ depends affinely on the design parameters x and, hence, closed-loop convex design specifications translate into convex constraints on x. Or, equivalently, by substitution of relations (2.13) and (2.15), the sets $\mathscr{S}_i$ are converted to convex sets for x.

Youla parameter $X(z)$ must be stable and therefore, stable basis functions $X_m(z)$ are required. In addition, the set of basis functions should be complete, that is: for $M \to \infty$, any stable transfer function $X(z)$ can be written as (2.15). In this monograph, the following basis functions are used:

$$X_m(z) = z^{1-m} \,, \tag{2.17}$$

which comply with the aforementioned requirements, while the reader is referred to [63] for alternatives.

2.5 Numerical Solution of Optimal Controller Design

The Youla parametrization translates the multi-objective control problem into a convex optimization problem in x. However, often this problem is not yet numerically tractable due to semi-infinite constraints that require evaluation on infinitely many frequencies. This section shows how to render such design problems numerically tractable by illustrating this transformation for the problem presented in Section 2.3.3:

$$\underset{x,\gamma_\mathrm{p},\gamma_\mathrm{np}}{\text{minimize}} \quad \gamma_\mathrm{p} + \alpha\gamma_\mathrm{np} \tag{2.18a}$$

$$\text{subject to} \quad \|H_\mathrm{np}(z,x)\|_\infty \le \gamma_\mathrm{np} \tag{2.18b}$$

$$W_l |H_\mathrm{p}(\omega,x)| \le \gamma_\mathrm{p} \,, \quad \forall \omega \in \Omega_l \,, \quad \forall l \in \mathscr{L} \,. \tag{2.18c}$$

Variable x is added as an argument in $H_\mathrm{np}(z,x)$ and $H_\mathrm{p}(z,x)$ to indicate relations (2.13) and (2.15). Note that internal closed-loop stability no longer involves a constraint, since the Youla parametrization translates internal stability into stability of $X(z)$, which is guaranteed by the stability of the basis functions $X_m(z)$.

Constraint (2.18b) is semi-infinite since it requires evaluating $|H_{np}(\omega, x)| \leq \gamma_{np}$ for all ω in $[0, \pi f_s]$. Similarly, constraints (2.18c) require evaluating an inequality on the continuous frequency intervals Ω_l.

2.5.1 Gridding

The most straightforward way to render optimization problem (2.18) numerically tractable is known as gridding: instead of evaluating the constraints on the continuous frequency intervals, they are only evaluated on a finite number of frequencies in these intervals. Let sets $\mathcal{G}_{np}$ and $\mathcal{G}_{p,l}$ comprise the frequency grid points considered in $[0, \pi f_s]$ and Ω_l, respectively, then problem (2.18) is reformulated as:

$$\underset{x, \gamma_p, \gamma_{np}}{\text{minimize}} \quad \gamma_p + \alpha \gamma_{np}$$

$$\text{subject to} \quad \left| \Phi_{1,np}(\omega) + \sum_{m=1}^{M} x_m \Phi_{2,np}(\omega) X_m(\omega) \Phi_{3,np}(\omega) \right| \leq \gamma_{np}, \quad \forall \omega \in \mathcal{G}_{np}$$

$$\left| \Phi_{1,p}(\omega) + \sum_{m=1}^{M} x_m \Phi_{2,p}(\omega) X_m(\omega) \Phi_{3,p}(\omega) \right| \leq \gamma_p, \quad \begin{array}{l} \forall \omega \in \mathcal{G}_{p,l}, \\ \forall l \in \mathcal{L}. \end{array}$$

Each of the amplitude constraints translates into a second-order cone constraint. Although care is required in selecting the frequency grid points, the resulting second-order cone problem (SOCP) is generally solved very efficiently with standard interior-point solvers, such as SDPT3 [141, 149]. This involves a computational complexity of $\mathcal{O}(n_{\mathcal{G}} M^2)$, where $n_{\mathcal{G}}$ denotes the total number of grid points. In principle, gridding does not require a parametric plant model, being able to deal with nonparametric FRF estimates of the plant at the gridding frequencies. However, it should be noted that obtaining a sufficiently fine frequency grid may invoke long identification experiments.

2.5.2 KYP and Generalized KYP Lemma

The mathematically more elegant way to render optimization problem (2.18) numerically tractable is to transform the semi-infinite constraints into linear matrix inequalities (LMIs) by application of the Kalman-Yakubovich-Popov (KYP) lemma [79, 113, 157] and the generalized KYP lemma [77, 123]. These transformations are detailed below, while more details on the (generalized) KYP lemma as well as the numerical solution of the resulting semi-definite programming problem (SDP) are provided in Appendix A.

As a consequence of the Youla parametrization, only the numerator coefficients of $H_p(z, x)$ and $H_{np}(z, x)$ depend on x, and this in an affine manner. Based on this property, state-space models for $H_p(z,x)$ and $H_{np}(z,x)$ can be derived where the state equation is independent of x:

$$H_p(z,x) = C_p(x)\,(zI-A)^{-1}B_p + D_p(x)\,,$$
$$H_{np}(z,x) = C_{np}(x)\,(zI-A)^{-1}B_{np} + D_{np}(x)\,,$$

while $C_p(x)$, $D_p(x)$, $C_{np}(x)$ and $D_{np}(x)$ are affine in x. For instance, the control canonical state-space form satisfies these requirements. The order of the closed-loop system $H(z,x)$, and hence, of $H_p(z,x)$ and $H_{np}(z,x)$ is denoted n.

The KYP lemma states that constraint (2.18b) is equivalent to the existence of a matrix $P_{np} \in \mathbf{S}_n$ that satisfies:

$$\begin{bmatrix} A^T P_{np} A - P_{np} & A^T P_{np} B_{np} & C_{np}(x)^T \\ B_{np}^T P_{np} A & B_{np}^T P_{np} B_{np} - \gamma_{np} & D_{np}(x)^T \\ C_{np}(x) & D_{np}(x) & -\gamma_{np} \end{bmatrix} \preceq 0\,. \tag{2.19}$$

Since relations $C_{np}(x)$ and $D_{np}(x)$ are affine, this matrix inequality corresponds to an LMI in x and P_{np}.

The generalized KYP lemma states that for each $l \in \mathscr{L}$, constraint (2.18c) is equivalent to the existence of matrices $P_{p,l} \in \mathbf{H}_n$ and $Q_{p,l} \in \mathbf{H}_n$ that satisfy the following set of LMIs:

$$\begin{bmatrix} A^T P_{p,l} A - P_{p,l} & A^T P_{p,l} B_p & C_p(x)^T \\ B_p^T P_{p,l} A & B_p^T P_{p,l} B_p - \gamma_p & D_p(x)^T \\ C_p(x) & D_p(x) & -\gamma_p \end{bmatrix} +$$
$$\begin{bmatrix} \eta_l A^T Q_{p,l} + \eta_l^H Q_{p,l} A + \zeta_l Q_{p,l} & \eta_l^H Q_{p,l} B_p & 0 \\ \eta_l B_p^T Q_{p,l} & 0 & 0 \\ 0 & 0 & 0 \end{bmatrix} \preceq 0\,, \tag{2.20a}$$
$$Q_{p,l} \succeq 0\,, \tag{2.20b}$$

where $\zeta_l = -2\cos(l\omega_p\delta T_s)$, $\eta_l = \exp(jl\omega_p T_s)$.

Relying on the aforementioned transformations, optimization problem (2.18) is transformed into the following, equivalent SDP:

$$\begin{aligned} \text{minimize}\quad & \gamma_p + \alpha\gamma_{np} \\ \text{subject to}\quad & \text{LMI (2.19)} \\ & \text{LMI (2.20a)}\,, \quad \forall l \in \mathscr{L} \\ & \text{LMI (2.20b)}\,, \quad \forall l \in \mathscr{L}\,, \end{aligned}$$

where the optimization variables are γ_p, γ_{np}, x, P_p, and $P_{p,l}$ and $Q_{p,l}$ for all $l \in \mathscr{L}$.

Solving this SDP with a standard interior-point solver like SDPT3 [141, 149] involves a computational complexity of $\mathcal{O}(n^6)$. On the other hand, as detailed in Appendix A, the solution approach proposed in [150] reduces the complexity to $\mathcal{O}(n^4)$, whereas the solver presented in [94] is even more efficient: $\mathcal{O}(n^3)$.

2.6   Note on Continuous-time and MIMO Control

All building blocks of the methodology remain valid in continuous time, except for the selected set (2.17) of basis functions for the Youla parameter $X(z)$. Stable and complete sets of continuous-time basis transfer functions are found in [63].

Moving from a SISO controller design to a MIMO design only requires minor and rather straightforward modifications to the methodology, but complicates the numerical solution of the resulting optimization problem.

No assumption is made concerning the dimension of exogenous inputs $w_i(k)$ and regulated outputs $v_i(k)$, except for $w_p(k)$, which is assumed scalar. If $w_p(k)$ is a vector signal, the methodology allows for two cases. In the first case, the direction e_l of each harmonic component is known, where e_l corresponds to a one-normed vector such that the contribution to $w_p(k)$ of the l'th harmonic is of the form

$$2W_l \sin(l\omega_p k T_s + \phi_l)\, e_l.$$

For instance, under this assumption, each constraint (2.18c) is replaced by

$$W_l |H_p(\omega,x)e_l| \leq \gamma_p, \quad \forall \omega \in \Omega_l.$$

In the alternative case, directions e_l are unknown, and to account for the worst-case directions, constraints (2.18c) are replaced by

$$W_l\, \sigma_{\max}\{H_p(\omega,x)\} \leq \gamma_p, \quad \forall \omega \in \Omega_l,$$

where $\sigma_{\max}\{X\}$ denotes the largest singular value of X. Neither of these cases compromises the convexity of the optimization problem nor its transformation into an SDP.

2.7   Conclusion

This chapter presents the general design methodology for controllers facing periodic input signals and elaborates on its fundamental concepts. The design methodology is multi-objective in nature, combining improved closed-loop periodic performance with additional design specifications. The multi-objective controller design is formulated in terms of the general control configuration and to incorporate good closed-loop periodic performance in the design, it is quantified by the periodic performance index, hereby explicitly accounting for period-time uncertainty. By application of the Youla parametrization, the design problem is transformed into a convex optimization problem, which is rendered numerically tractable by application of gridding or the (generalized) KYP lemma.

Chapter 3
Application to Feedforward Control

3.1 Introduction

3.1.1 State of the Art

To meet the continual quest for higher tracking accuracy in engineering practice, feedforward controllers have become an essential complement to feedback control. The ideal feedforward controller inverts the (closed-loop) system and as it yields perfect tracking for any reference input, specialized feedforward design for periodic inputs seems superfluous. However, the ideal feedforward controller suffers from two deficiencies, hereby raising applications where exploiting the input periodicity is beneficial.

First, the practical implementation of the ideal feedforward controller is impeded by nonminimum-phase zeros of the plant. Such zeros are common in discrete-time plant models, where they originate from noncollocated control [115, 147, 168] or stem from the zero-order hold discretization of a continuous-time system [8]. As inverting nonminimum-phase zeros translates into poles outside the unit circle, the causal[1] implementation of the ideal feedforward controller is unstable. Although stable, its noncausal implementation [39, 73] is neither appealing in practice since it requires infinite preview time: the controller needs all future values of the reference trajectory to compute its current output. In addition, the noncausal ideal feedforward controller requires a certain pre-actuation time, building up control signal before the plant output follows the reference trajectory[2].

The second deficiency of the ideal feedforward controller is related to model uncertainty: it inverts the discrete-time plant model and hereby inherently assumes

---

[1] For ease of explanation, the plant is assumed to have zero relative degree.

[2] The working principle of the noncausal ideal feedforward is to build up the initial conditions for the plant such that the unstable component of the causal free and forced response cancel. This initial condition build-up takes infinite time if it has to comply with the zero dynamics of the plant, such that meanwhile no output it generated. If this requirement is relaxed, finite pre-actuation time suffices [115].

G. Pipeleers et al.: Optimal Linear Controller Design for Periodic Inputs, LNCIS 394, pp. 23–42.
springerlink.com   © Springer-Verlag Berlin Heidelberg 2009

high model accuracy. However, in many applications the validity of this assumption is limited, and the tracking performance of the ideal feedforward controller is very sensitive to model uncertainties [38].

Although the literature reveals active research on designing feedforward controllers for nonminimum-phase systems [18, 29, 54, 58, 115, 142, 147, 166, 167], only few contributions deal with specializing the feedforward controller for periodic inputs in these applications. However, for periodic inputs the infinite preview time requirement of the noncausal ideal feedforward controller reduces to one period of preview, which implies that after one period, perfect tracking is possible. Tomizuka *et al.* [143] propose a feedforward design that splits up the periodic input in its harmonic components and pre-compensates for each harmonic the phase and amplitude distortion of the plant. The implementation of this controller is cumbersome, and Walgama and Sternby [152] propose a feedforward design much easier to implement. The feedforward controller is designed as a finite impulse response (FIR) filter that inverts the plant only at the input harmonics, yielding perfect asymptotic tracking of periodic reference inputs. This FIR filter design, briefly reviewed in Section 3.2.1, is analytical (the filter coefficients are computed as the solution of a set of linear equations), assumes perfect knowledge of the input period and cannot cope with plant uncertainty.

Although robust feedforward control gains increasing attention [44, 52, 89, 126], specializing the robust feedforward controller design for periodic inputs is not yet addressed in the literature. However, specializing the robust feedforward controller design for periodic inputs is advantageous, as in this case, robust performance is only affected by the plant uncertainty around the input harmonics.

3.1.2   Contribution

This chapter applies the general methodology of Chapter 2 to design a feedforward controller for periodic inputs to a discrete-time SISO LTI system. Hereby, the design methodology extends the FIR filter design of Walgama and Sternby [152] with the following advantages:

Period-time Uncertainty: Instead of enforcing perfect tracking for the nominal input period, the primal objective in the design methodology is minimizing the periodic performance index, which explicitly accounts for period-time uncertainty.

Multi-objective Control: Whereas in [152] all design freedom is attributed to eliminating the periodic tracking error, the developed design methodology allows accounting for a variety of additional design specifications, such as transient response, control effort, etc.

Plant Uncertainty: Contrary to [152], the design methodology allows accounting for plant uncertainty.

3.1.3 Outline

While Section 3.2 reviews the feedforward design approach of Walgama and Sternby [152], Section 3.3 applies the developed methodology to design a feedforward controller and elaborates on the corresponding general control configuration, Youla parametrization and optimal design. Section 3.4 illustrates the potential of the design methodology by numerical results.

3.2 Background

This section surveys commonly used feedforward control configurations (Section 3.2.1) and summarizes the design approach of Walgama and Sternby [152] (Section 3.2.2).

3.2.1 Control Configuration

Figure 3.1 shows common feedforward control configurations. Feedforward controller $K_{FF}(z)$ is added to improve tracking of reference trajectory $r(k)$, Figures 3.1(a), 3.1(c) and 3.1(d); or to improve attenuation of measurable disturbance $d(k)$ with known effect $G_d(z)$ on the plant output $\eta(k)$, Figures 3.1(b) and 3.1(e). $K_{FF}(z)$ is either applied to the open-loop plant $G(z)$, Figures 3.1(a) and 3.1(b); or combined with a feedback controller $K_o(z)$, Figures 3.1(c), 3.1(d) and 3.1(e). In the latter case, $K_o(z)$ is assumed to be designed *a priori* and the corresponding closed-loop sensitivity and complementary sensitivity are denoted by $S_o(z)$ and $T_o(z)$, respectively:

$$S_o(z) = \frac{1}{1 + K_o(z)G(z)} , \qquad T_o(z) = \frac{K_o(z)G(z)}{1 + K_o(z)G(z)} .$$

$K_o(z)$ must yield an internally stable feedback system and is hence indispensable if $G(z)$ is unstable.

For the sake of unified treatment, the control architectures of Figure 3.1 are converted to the more general configuration of Figure 3.2, where these conversions are detailed in Table 3.1. The exogenous input is considered periodic and specified according to Section 2.2.2, and is hence indicated by $w_p(k)$. This input is directly accessible to the feedforward controller $K_{FF}(z)$, which converts it to the control signal $u(k)$. Stable systems $P_p(z)$ and $P_{pu}(z)$ respectively relate $w_p(k)$ and $u(k)$ to the regulated output $v_p(k)$. The overall (closed-loop) system $H_p(z)$ from $w_p(k)$ to $v_p(k)$ is given by

$$H_p(z) = P_p(z) + P_{pu}(z)K_{FF}(z) , \qquad (3.1)$$

and stability of $H_p(z)$ requires a stable feedforward controller.

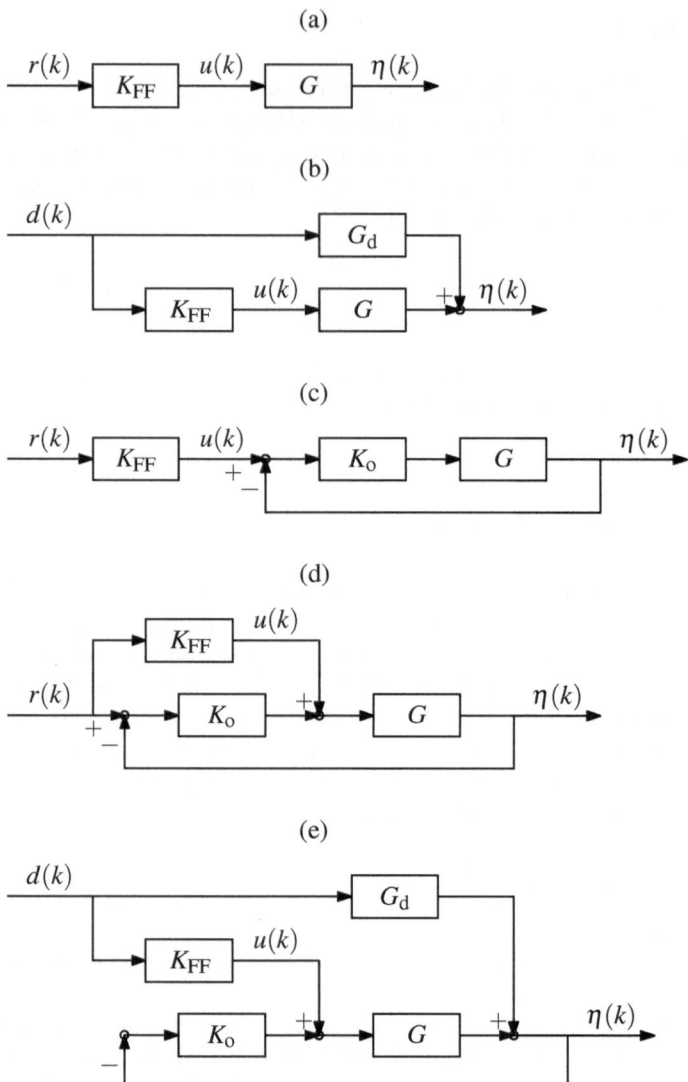

Fig. 3.1 Common feedforward control configurations: feedforward controller K_{FF} is designed for plant G to improve tracking of reference input $r(k)$, (a), (c) and (d); or to improve attenuation of measurable disturbance $d(k)$ with known effect G_d on the plant output $\eta(k)$, (b) and (e). K_{FF} is either applied in open loop, (a) and (b); or combined with a feedback controller K_o, (c), (d) and (e).

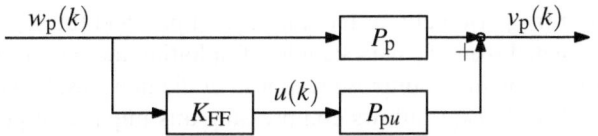

Fig. 3.2 More general feedforward control configuration.

Table 3.1 Conversion of the feedforward control configurations of Figure 3.1 to the more general one of Figure 3.2.

configuration	$w_p(k)$	$v_p(k)$	$P_p(z)$	$P_{pu}(z)$
Figure 3.1(a)	$r(k)$	$r(k) - \eta(k)$	1	$-G(z)$
Figure 3.1(b)	$d(k)$	$\eta(k)$	$G_d(z)$	$G(z)$
Figure 3.1(c)	$r(k)$	$r(k) - \eta(k)$	1	$-T_o(z)$
Figure 3.1(d)	$r(k)$	$r(k) - \eta(k)$	$S_o(z)$	$-S_o(z)G(z)$
Figure 3.1(e)	$d(k)$	$\eta(k)$	$S_o(z)G_d(z)$	$S_o(z)G(z)$

3.2.2 Current Design Approach

Walgama and Sternby [152] design $K_{FF}(z)$ as a FIR filter of length n_Λ:

$$K_{FF}(z) = \sum_{m=1}^{n_\Lambda} k_{FF,m} z^{1-m} , \qquad (3.2)$$

where n_Λ corresponds to the order of signal generator $\Lambda(z)$, (2.5), and is given by (2.6). To guarantee perfect asymptotic tracking/rejection of $w_p(k)$ for nominal period T_p, the n_Λ filter coefficients $k_{FF,m}$ are computed such that

$$H_p(l\omega_p) = P_p(l\omega_p) + P_{pu}(l\omega_p)K_{FF}(l\omega_p) = 0 , \quad \forall l \in \mathscr{L} . \qquad (3.3)$$

The corresponding FIR filter exists provided that $P_p(z)^{-1}P_{pu}(z)$ has no zeros coinciding with harmonics in $\mathscr{L}$, and it is unique if $P_p(z)$ and $P_{pu}(z)$ have no common zeros coinciding with a harmonic in $\mathscr{L}$.

There are four ways to compute the filter coefficients $k_{FF,m}$ such that (3.3) is satisfied:

1. By solving a diophantine equation, zeros are enforced in the closed-loop transfer function $H_p(z)$ on the harmonics $l \in \mathscr{L}$, that is: on the poles of $\Lambda(z)$; see [152].
2. The filter coefficients are computed by solving the set of linear equations corresponding to (3.3); see [152]. Each of the constraints (3.3) translates into two real constraints, except for $l = 0$ and $l = T_p f_s/2$, which yield only one constraint since at 0 Hz and $f_s/2$ the FRF of a discrete-time system is real.

3. On account of the periodicity of input $w_p(k)$, all past and future values of $w_p(k)$ can be constructed from n_Λ data samples. Exploiting these relations in the convolution of $w_p(k)$ with the impulse response of the noncausal ideal feedforward controller $-P_{pu}(z)^{-1}P_p(z)$ allows compressing this impulse response to a finite and causal series of length n_Λ; see [152].

4. Within the theory of output regulation (see Appendix B for an introduction and [117] for an in-depth treatment), the FIR feedforward controller corresponds to a particular state-feedback controller. While in output regulation theory, a state-feedback controller generally feeds back the states of both plant and signal generator $\Lambda(z)$, for a stable plant, the controller can be designed to only feed back the signal generator states. As the states of $\Lambda(z)$ are linearly related to the n_Λ last samples of $w_p(k)$, this particular state-feedback controller corresponds to the FIR feedforward controller of [152].

Whereas solution strategies 1, 3, and 4 require parametric models for $P_p(z)$ and $P_{pu}(z)$, solution strategy 2 suffices with nonparametric FRF estimates of these systems at the input harmonics. However, uncertainty on these FRF estimates cannot be accounted for, and the same holds for uncertainty on the input period.

3.3   Application of the Design Methodology

This section applies the general methodology of Chapter 2 to design a feedforward controller for periodic inputs to a discrete-time SISO LTI system and elaborates on the corresponding general control configuration (Section 3.3.1), Youla parametrization (Section 3.3.2) and optimal design (Section 3.3.3). Section 3.3.4 renders the optimal feedforward controller design robust for unstructured plant uncertainty[3].

3.3.1   General Control Configuration

Figure 3.3 illustrates how to transform the feedforward control configuration of Figure 3.2 into the general control configuration (Figure 2.1). This yields the following generalized plant:

$$\begin{bmatrix} v_p(k) \\ y(k) \end{bmatrix} = \underbrace{\begin{bmatrix} P_p(q) & P_{pu}(q) \\ 1 & 0 \end{bmatrix}}_{P(q)} \begin{bmatrix} w_p(k) \\ u(k) \end{bmatrix} , \tag{3.4a}$$

and hence,

$$P_{yp}(q) = 1 , \tag{3.4b}$$

$$P_{yu}(q) = 0 . \tag{3.4c}$$

---

[3] Chapters 3 and 4 focus on rendering the developed design methodology robust for plant uncertainty, while Chapters 5 and 6 emphasize the methodology's multi-objective nature, trading off closed-loop periodic performance against conflicting design objectives.

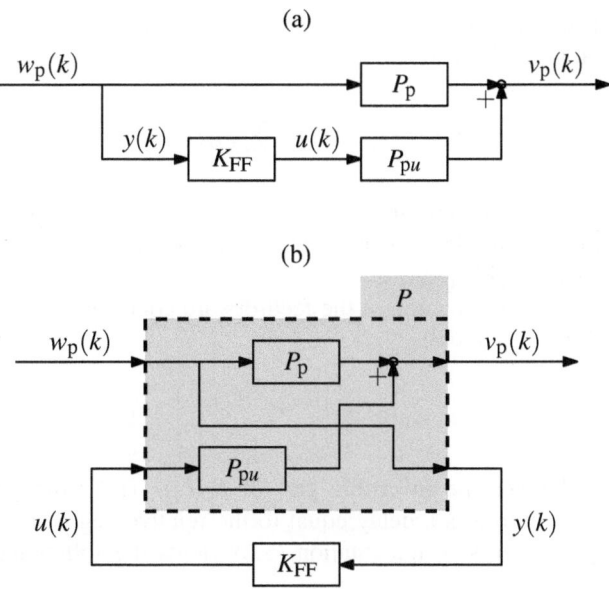

Fig. 3.3 Conversion of feedforward control configuration (a) to the general control configuration (b).

Accounting for additional design specifications requires extending $P(z)$ with complementary exogenous inputs and regulated outputs.

3.3.2 Youla Parametrization

By proper plant augmentation, the Youla parametrization translates the controller design into the design of Youla parameter $X(z)$, which relates affinely to the closed-loop system $H(z)$. The augmented plant $\widetilde{P}(z)$ must be stable and satisfy $\widetilde{P}_{\tilde{y}\tilde{u}}(z) = 0$, and since for a feedforward controller design, plant $P(z)$ (3.4) already satisfies these conditions, there is no need for plant augmentation:

$$K_{\mathrm{nom}}(z) = 0 \,,$$

and a proper augmentation of this controller is given by

$$\begin{bmatrix} u(k) \\ \tilde{y}(k) \end{bmatrix} = \underbrace{\begin{bmatrix} 0 & 1 \\ 1 & 0 \end{bmatrix}}_{\widetilde{K}_{\mathrm{nom}}(q)} \begin{bmatrix} y(k) \\ \tilde{u}(k) \end{bmatrix} \,.$$

This way, the Youla parameter $X(z)$ acts as a feedforward controller, and application of parametrization (2.17) yields

$$K_{FF}(z) = X(z) = \sum_{m=1}^{M} x_m z^{1-m} . \qquad (3.5)$$

The design parameters x_m are grouped in the vector $x \in \mathbf{R}_M$, Equation 2.16, and for $M \geq n_\Lambda$, this parametrization encompasses the feedforward controller (3.2) of Walgama and Sternby [152].

An alternative parametrization of the feedforward controller is obtained by augmenting $K_{nom} = 0$ as follows[4]:

$$\widetilde{K}_{nom}(z) = \begin{bmatrix} 0 & P_{pu,-}(z)^{-1} \\ 1 & 0 \end{bmatrix} , \qquad (3.6)$$

where $P_{pu,-}(z)$ denotes the invertible part of $P_{pu}(z)$. The remaining noninvertible part $P_{pu,+}(z)$ comprises a delay equal to the relative degree of $P_{pu}(z)$ and its nonminimum-phase zeros. Augmentation (3.6) yields the following feedforward controller:

$$K_{FF}(z) = P_{pu,-}(z)^{-1}X(z) = P_{pu,-}(z)^{-1} \sum_{m=1}^{M} x_m z^{1-m} , \qquad (3.7)$$

and this alternative parametrization is of particular interest for configurations (a) and (c) of Figure 3.1: for these configurations it yields a FIR closed-loop system:

$$H_p(z) = 1 + P_{pu,+}(z)X(z) ,$$

and the design of x only depends on the noninvertible part of $P_{pu}(z)$.

3.3.3   Optimal Design

Good steady-state closed-loop attenuation/tracking of periodic input $w_p(k)$ is the main objective in the feedforward controller design. To quantify this objective, the 2-norm based periodic performance index $\gamma_{p,2}$ (2.8) is suggested, since the designer often has a good idea about the spectrum of $w_p(k)$ as it corresponds to a reference trajectory or measurable disturbance. Period-time uncertainty is explicitly accounted for in the definition of $\gamma_{p,2}$.

Without additional design specifications, the FIR filter coefficients x_m, are computed as the solution of the following optimization problem:

---

[4] In fact, any two-by-two system with zero diagonal elements, and stable and invertible off-diagonal elements is allowed for $\widetilde{K}_{nom}(z)$, see Section 2.4.

$$\underset{x,\gamma_{p,2},V_l}{\text{minimize}} \quad \gamma_{p,2} \tag{3.8a}$$

$$\text{subject to} \quad \sqrt{\sum_{l\in\mathscr{L}} V_l^2} \leq \gamma_{p,2} \tag{3.8b}$$

$$W_l |H_p(\omega)| \leq V_l, \quad \forall \omega \in \Omega_l, \quad \forall l \in \mathscr{L}. \tag{3.8c}$$

This design problem can be supplemented with additional design specifications, such as constraints on the transient response of $H_p(z)$, reduced actuator effort, limited effect of measurement noise (if $w_p(k)$ corresponds to a measured disturbance), etc. (see e.g. [13] for an overview). In view of the discussion in Section 2.3.3, if the relative degree of $P_{pu}(z)$ is larger than the one of $P_p(z)$, a causal feedforward controller is bound to a trade-off between periodic and nonperiodic performance. The analysis of this trade-off is similar to the elaboration in Section 6.5.2, and therefore not further discussed here. Instead, it is illustrated below how to render optimization problem (3.8) robust for plant uncertainty.

3.3.4 *Optimal Robust Design for Plant Uncertainty*

This section renders optimization problem (3.8) robust for multiplicative unstructured plant uncertainty. This robustification is elaborated for the configuration of Figure 3.1(a), as for this configuration the mathematics are most intuitive. Appendix C elaborates the more general approach to the robust feedforward controller design, which relies on the structured singular value [106, 131, 163]. This approach applies to all configurations of Figure 3.1, and allows extending the robust feedforward controller design to structured plant uncertainty, although this generally involves conservatism [44, 52].

Instead of accounting for the nominal plant $G(z)$ solely, a robust feedforward controller performs well for all potential plant models $G_\Delta(z)$ of the form (2.2). According to control configuration Figure 3.1(a), the corresponding set of potential closed-loop systems $H_{p,\Delta}(z)$ is given by

$$
\begin{aligned}
H_{p,\Delta}(z) &= 1 - K_{FF}(z)G_\Delta(z), & \Delta(z) \in \Delta, \\
&= 1 - K_{FF}(z)G(z)\left[1 + W_G(z)\Delta(z)\right], & \Delta(z) \in \Delta,
\end{aligned}
$$

where Δ is given by (2.2b).

On count of the stability assumption on $G(z)$, $W_G(z)$ and $\Delta(z)$, all potential plant models $G_\Delta(z)$ are stable. Consequently, in a feedforward controller design robust closed-loop stability is equivalent to nominal closed-loop stability and only requires stability of $K_{FF}(z)$. Hence, the major issue in a robust feedforward controller design is robust closed-loop performance, which requires $K_{FF}(z)$ to yield good performance for all potential plant models $G_\Delta(z)$. To obtain good robust periodic performance, for each harmonic $l \in \mathscr{L}$, constraint (3.8c) is replaced by

$$W_l |H_{\mathrm{p},\Delta}(\omega)|_{\mathrm{wc}} \le V_{l,\mathrm{wc}}, \quad \forall \omega \in \Omega_l, \tag{3.9a}$$

where

$$
\begin{aligned}
|H_{\mathrm{p},\Delta}(\omega)|_{\mathrm{wc}} &= \max_{|\Delta(\omega)| \le 1} \left\{ |H_{\mathrm{p},\Delta}(\omega)| \right\}, \\
&= \max_{|\Delta(\omega)| \le 1} \left\{ \left|1 - K_{\mathrm{FF}}(\omega)G(\omega) - K_{\mathrm{FF}}(\omega)G(\omega)W_G(\omega)\Delta(\omega)\right| \right\}.
\end{aligned}
$$

The complex scalar $\Delta(\omega)$ that maximizes the right-hand side, has modulus $|\Delta(\omega)| = 1$ and its phase aligns $[K_{\mathrm{FF}}(\omega)G(\omega)W_G(\omega)\Delta(\omega)]$ opposite to $[1 - K_{\mathrm{FF}}(\omega)G(\omega)]$. This way,

$$|H_{\mathrm{p},\Delta}(\omega)|_{\mathrm{wc}} = |1 - K_{\mathrm{FF}}(\omega)G(\omega)| + |K_{\mathrm{FF}}(\omega)G(\omega)W_G(\omega)|, \tag{3.9b}$$

and the robust counterpart of (3.8) amounts to

$$\underset{x,\gamma_{\mathrm{p},2},V_{l,\mathrm{wc}}}{\text{minimize}} \quad \gamma_{\mathrm{p},2} \tag{3.10a}$$

$$\text{subject to} \quad \sqrt{\sum_{l \in \mathscr{L}} V_{l,\mathrm{wc}}^2} \le \gamma_{\mathrm{p},2} \tag{3.10b}$$

$$W_l |H_{\mathrm{p},\Delta}(\omega)|_{\mathrm{wc}} \le V_{l,\mathrm{wc}}, \quad \forall \omega \in \Omega_l, \quad \forall l \in \mathscr{L}. \tag{3.10c}$$

After substituting relation (3.5) or (3.7) in Equation 3.9b, constraints (3.10c) correspond to convex semi-infinite constraints in x. These constraints cannot be converted into LMIs by the generalized KYP lemma[5], and hence, gridding is required to render (3.10) numerically tractable.

3.4 Numerical Results

This section illustrates the potential of the design methodology of Chapter 2 for a feedforward control problem by numerical results, where the simulation example is presented in Section 3.4.1. First, optimal design problem (3.8) is considered and the solution is compared with the design methodology of Walgama and Sternby [152]. Subsequently, Section 3.4.3 discusses the robust feedforward controller design for unstructured plant uncertainty.

---

[5] The generalized KYP lemma does apply to:

$$W_l \sqrt{\left|1 - K_{\mathrm{FF}}(\omega)G(\omega)\right|^2 + \left|K_{\mathrm{FF}}(\omega)G(\omega)W_G(\omega)\right|^2} \le V_{l,\mathrm{wc}}, \quad \forall \omega \in \Omega_l,$$

which is closely related to (3.9), since for two complex numbers α and β, the equivalence of norms implies:

$$\sqrt{|\alpha|^2 + |\beta|^2} \le |\alpha| + |\beta| \le \sqrt{2}\sqrt{|\alpha|^2 + |\beta|^2}.$$

3.4.1 Simulation Example

According to Figure 3.1(a), the feedforward controller is designed to improve the open-loop tracking of a periodic reference trajectory. The simulation is executed at $f_s = 1$ kHz and reference input $w_p(k)$ corresponds to the periodic extension of the signal shown in Figure 3.4(a). The nominal period $T_p = 0.05$ s comprises $N = 50$ sample periods, and yields $f_p = 20$ Hz. However, the period is determined by an external process and may deviate from its nominal value with one sample period, invoking $\delta = 2\%$. Figure 3.4(b) shows the weights W_l used in the feedforward design, which correspond to the amplitude spectrum of $w_p(k)$ divided by its rms value. The set $\mathscr{L}$ of harmonics in $w_p(k)$ comprises 0 Hz and all odd harmonics, yielding $n_{\mathscr{L}} = 14$.

Youla parametrization (3.7) is adopted:

$$K_{FF}(z) = G_-(z)^{-1}X(z) , \qquad (3.11)$$

whereby the closed-loop transfer function $H_p(z)$, and hence, the design of x depends solely on the noninvertible part $G_+(z)$ of the plant $G(z)$:

$$H_p(z) = 1 - G(z)K_{FF}(z) = 1 - G_+(z)X(z) . \qquad (3.12)$$

As specializing the feedforward controller design for periodic inputs is most relevant for nonminimum-phase systems, $G_+(z)$ is chosen here to comprise a nonminimum-phase zero $z = 1.05$ and one sample delay, corresponding to $G(z)$ having relative degree one:

$$G_+(z) = \frac{-20z + 21}{z^2} .$$

Figure 3.5(a) shows the FRF of this system.

In Section 3.4.3, multiplicative unstructured uncertainty on $G(z)$ is considered, yielding a set of potential plant models $G_\Delta(z)$ of the form (2.2), where $|W_G(\omega)|$ is shown in Figure 3.5(b). This way, all potential closed-loop transfer functions $H_{p,\Delta}(z)$ are of the form

$$H_{p,\Delta}(z) = 1 - G_+(z)\left[1 + W_G(z)\Delta(z)\right]X(z) .$$

Length M of the FIR Youla parameter $X(z)$ is bounded by $M \leq N - 2 = 48$, such that the transient response of $H_p(z)$ is restricted to one period.

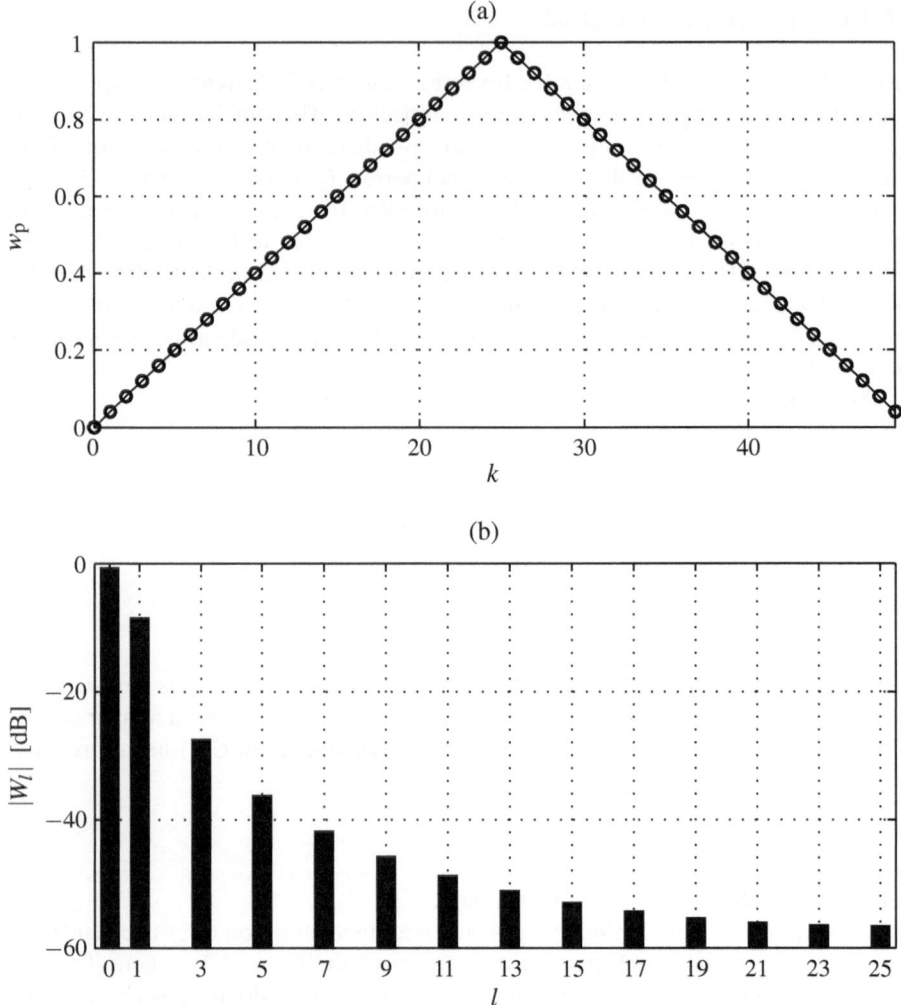

Fig. 3.4 (a) Periodic input signal $w_p(k)$; and (b) the corresponding weights W_l for the harmonics l, used in the optimal feedforward controller design.

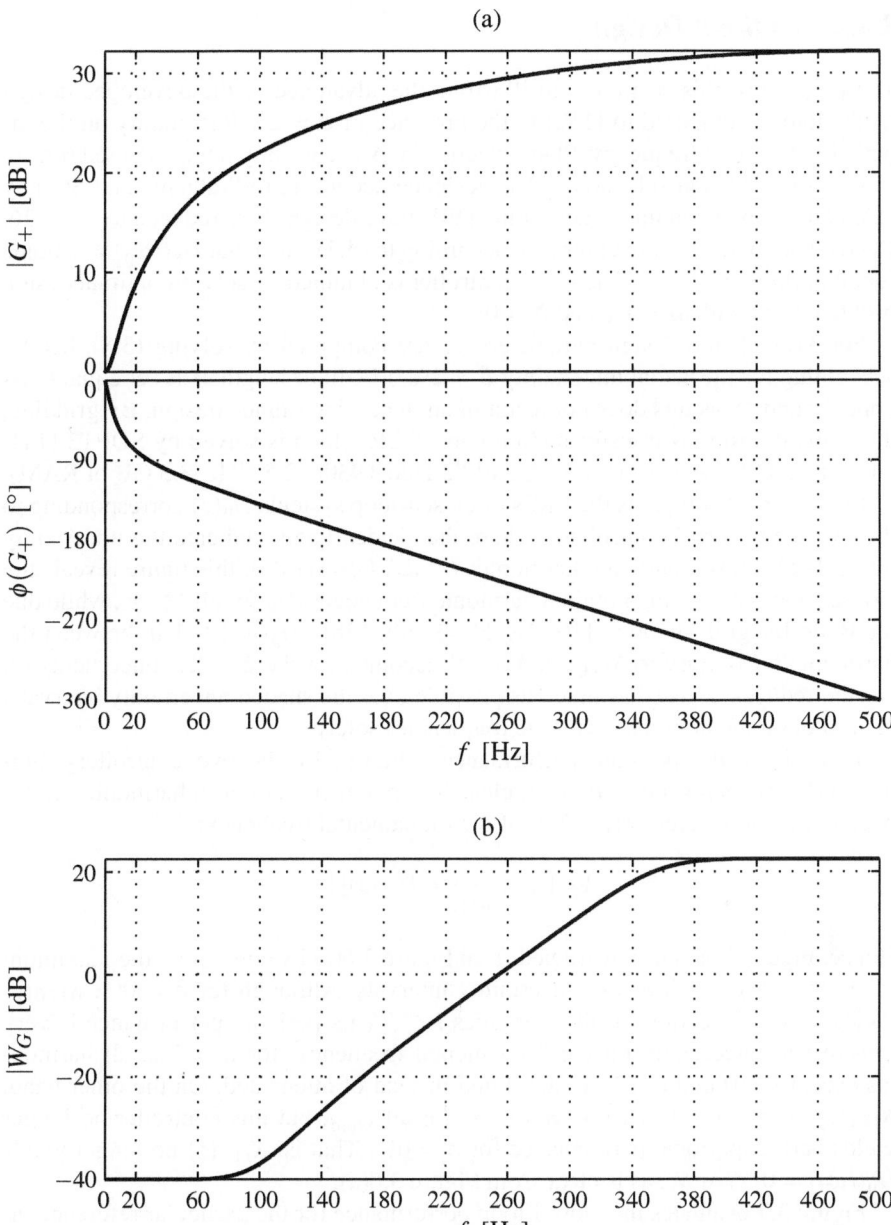

Fig. 3.5 (a) FRF of noninvertible part $G_+(z)$ of the plant $G(z)$; and (b) weight $|W_G(\omega)|$ of the multiplicative unstructured plant uncertainty.

3.4.2   Optimal Design

The purpose of this section is to illustrate the advantage of the developed design methodology compared to [152] in the presence of $\delta = 2\%$ uncertainty on the input's fundamental frequency. Plant uncertainty Δ is currently not accounted for.

Feedforward controller $K_{FF0}(z)$ is designed according to Walgama and Sternby [152] and to avoid an under-determined FIR filter design, M is reduced to $n_\Lambda = 26$. Design parameters x_m are computed according to (3.3), such that $H_p(l\omega_p) = 0$ holds for all harmonics $l \in \mathscr{L}$. The same controller is obtained by solving optimal design problem (3.8) with $M = n_\Lambda$ and $\delta = 0\%$.

For $K_{FF1}(z)$, the design parameters x_m are computed by solving (3.8), hereby accounting for the actual uncertainty $\delta = 2\%$. FIR filter length M is set equal to its upper bound 48 as this does not result in an under-determined design. By gridding, the optimal design is transformed into an SOCP, which is solved by SDPT3 [141, 149] in 1.3 CPU seconds (Intel® Core™2 Duo T9300, 2.5 GHz, 3.5 GB of RAM).

Figure 3.6(a) compares the FRFs of closed-loop systems $H_p(z)$ corresponding to the two feedforward controllers, where the shaded bands indicate the uncertainty intervals Ω_l (2.4) around the harmonics $l \in \mathscr{L}$. For $K_{FF0}(z)$, this figure reveals the closed-loop zeros on the nominal harmonic frequencies lf_p for all $l \in \mathscr{L}$, while due the Bode Integral Theorem [10, 21, 22, 49, 69, 138], $|H_p(\omega)| > 1$ in between the harmonics[6]. Contrary to $K_{FF0}(z)$, $K_{FF1}(z)$ accounts for the $\delta = 2\%$ uncertainty on f_p, and reduces $|H_p(\omega)|$ as much as possible over the shaded uncertainty intervals, instead of on the nominal harmonic frequencies solely.

To evaluate the periodic performance achieved by the two controllers, Figure 3.6(b) compares the worst-case closed-loop reduction of each harmonic $l \in \mathscr{L}$, over all potential values $\omega_{p,\delta}$ (2.3) of the fundamental frequency:

$$V_l/W_l = \max_{\omega \in \Omega_l} \left\{ |H_p(\omega)| \right\} .$$

Hence, Figure 3.6(b) is constructed from Figure 3.6(a) by computing the maximum of $|H_p(\omega)|$ over each of the uncertainty intervals. Although for $\delta = 0\%$, $K_{FF0}(z)$ yields perfect rejection of all harmonics $l \in \mathscr{L}$, its periodic performance is very sensitive to uncertainty on the fundamental frequency: for $\delta = 2\%$, all harmonics except $l = 0$ and $l = 1$ are amplified instead of attenuated. On the other hand, $K_{FF1}(z)$ attenuates all harmonics $l \in \mathscr{L}$ for all $\omega_{p,\delta}$, but this controller no longer yields perfect periodic performance for $\delta = 0\%$. That is: $K_{FF1}(z)$ no longer yields $H_p(l\omega_p) = 0$, $\forall l \in \mathscr{L}$, as is clear from Figure 3.6(a).

Figure 3.7 evaluates the closed-loop performance for the particular reference input $w_p(k)$ of Figure 3.4, by showing $\text{rms}(v_p(k))/\text{rms}(w_p(k))$ as a function of $f_{p,\delta}$. As the zero'th and first harmonic are dominant in $w_p(k)$, see Figure 3.4(b), $K_{FF0}(z)$ still yields an overall closed-loop reduction of $w_p(k)$ for all possible fundamental

---

[6] For this particular example, $|H_p(\omega)|$ becomes rather large in between the harmonics as the solution strategy of [152] places a closed-loop nonminimum-phase zero at $z = 15.97$, which increases the right-hand side of (2.11).

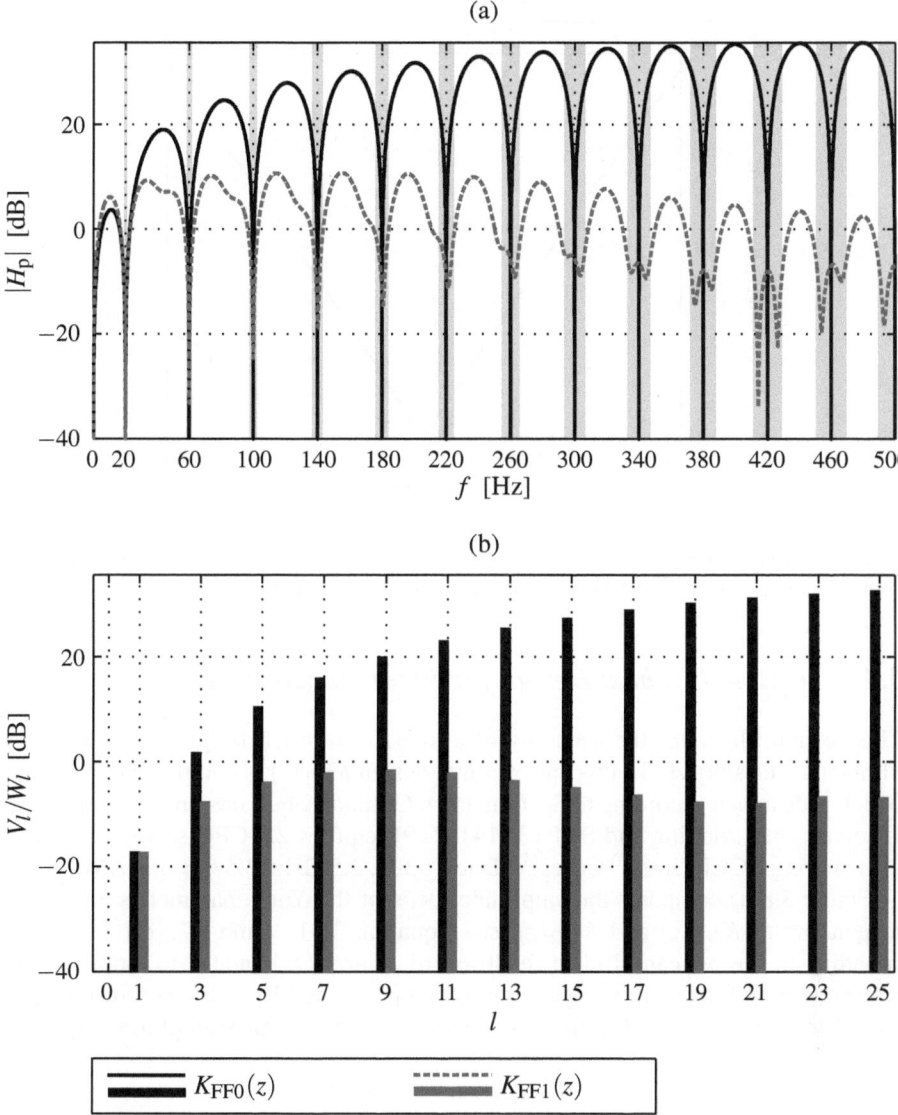

Fig. 3.6 Evaluation of feedforward controllers $K_{FF0}(z)$ [152] and $K_{FF1}(z)$ (optimal design): (a) amplitude FRF of the closed-loop transfer function $H_p(z)$; and (b) $V_l/W_l = \max_{\omega \in \Omega_l}\{|H_p(\omega)|\}$.

frequencies $f_{p,\delta}$. Compared to $K_{FF0}(z)$, the performance of $K_{FF1}(z)$ for $w_p(k)$ is significantly less sensitive to δ, where the price for this improved robust performance is moderate: instead of yielding $\mathrm{rms}(v_p(k)) = 0$ for $\delta = 0\%$, 0.4% of $\mathrm{rms}(w_p(k))$ remains.

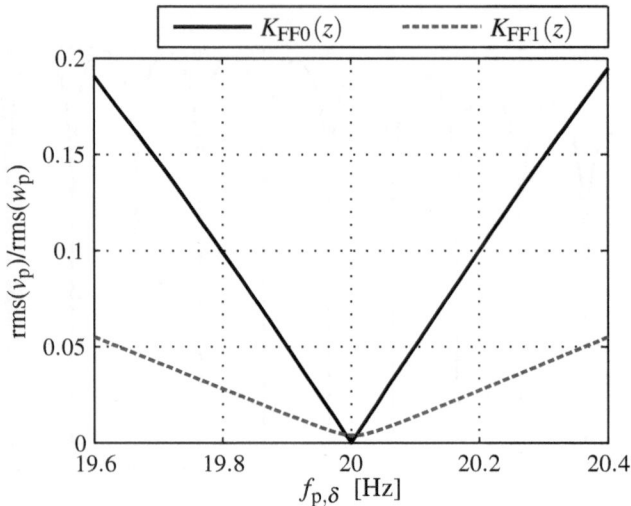

Fig. 3.7 Closed-loop rms reduction of the considered input $w_p(k)$ achieved by feedforward controllers $K_{FF0}(z)$ [152] and $K_{FF1}(z)$ (optimal design) as a function of $f_{p,\delta}$.

3.4.3 Optimal Robust Design for Plant Uncertainty

This section illustrates the necessity of a robust controller design in the presence of plant uncertainty Δ. To this end, optimal design $K_{FF1}(z)$ is compared to $K_{FF2}(z)$, which is designed according to Section 3.3.4. Optimal robust design problem (3.10) is solved with gridding and SDPT3 [141, 149] requires 2.9 CPU seconds to solve the resulting SOCP (Intel® Core™2 Duo T9300, 2.5 GHz, 3.5 GB of RAM).

Figure 3.8(a) compares the amplitude FRFs of the Youla parameters $X(z)$ corresponding to $K_{FF1}(z)$ and $K_{FF2}(z)$, see Equation 3.11, while $|G_+(\omega)^{-1}|$, corresponding to the noncausal ideal feedforward controller, is added to facilitate the interpretation of the results. Figure 3.8(b) compares the FRFs of the corresponding closed-loop systems $H_{p,\Delta}(z)$: the thin lines indicate the nominal amplitude $|H_p(\omega)|$:

$$|H_p(\omega)| = \left|1 - G_+(\omega)X(\omega)\right|, \qquad (3.13)$$

while the thick lines correspond to the worst-case amplitude $|H_{p,\Delta}(\omega)|_{wc}$ (3.9b):

$$|H_{p,\Delta}(\omega)|_{wc} = \left|1 - G_+(\omega)X(\omega)\right| + \left|W_G(\omega)G_+(\omega)X(\omega)\right|. \qquad (3.14)$$

For $K_{FF1}(z)$, the thin curve corresponds to the result shown in Figure 3.6(a).

Figure 3.9(a) evaluates the periodic performance achieved by the feedforward controllers for the nominal plant model, showing

$$V_l/W_l = \max_{\omega \in \Omega_l} \left\{|H_p(\omega)|\right\},$$

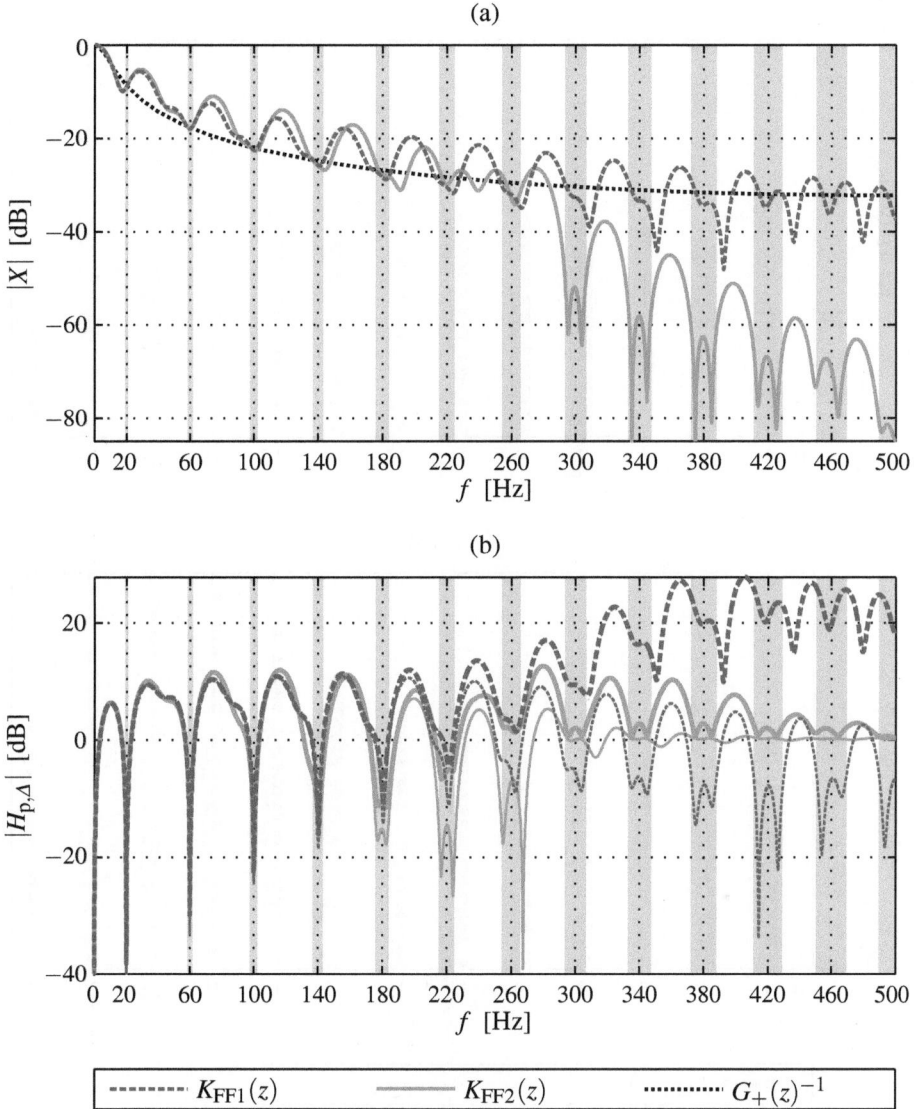

Fig. 3.8 Evaluation of feedforward controllers $K_{FF1}(z)$ (optimal design) and $K_{FF2}(z)$ (optimal robust design): (a) amplitude FRF of Youla parameter $X(z)$; and (b) amplitude FRF of the closed-loop transfer function $H_{p,\Delta}(z)$, where the thin and thick line respectively indicate the nominal amplitude $|H_p(\omega)|$ and worst-case amplitude $|H_{p,\Delta}(\omega)|_{wc}$.

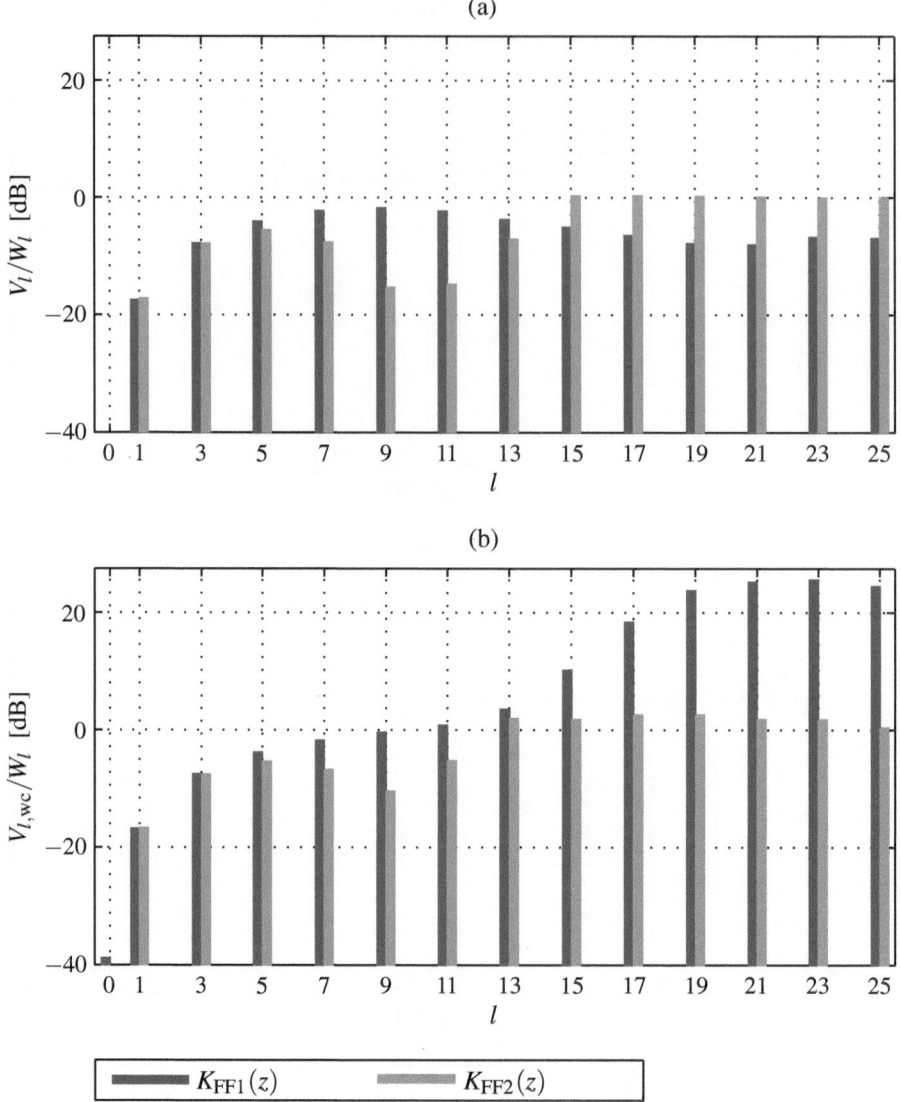

Fig. 3.9 Evaluation of feedforward controllers $K_{\mathrm{FF1}}(z)$ (optimal design) and $K_{\mathrm{FF2}}(z)$ (optimal robust design): (a) $V_l/W_l = \max_{\omega \in \Omega_l} \{|H_{\mathrm{p}}(\omega)|\}$; and (b) $V_{l,\mathrm{wc}}/W_l = \max_{\omega \in \Omega_l} \{|H_{\mathrm{p},\Delta}(\omega)|_{\mathrm{wc}}\}$.

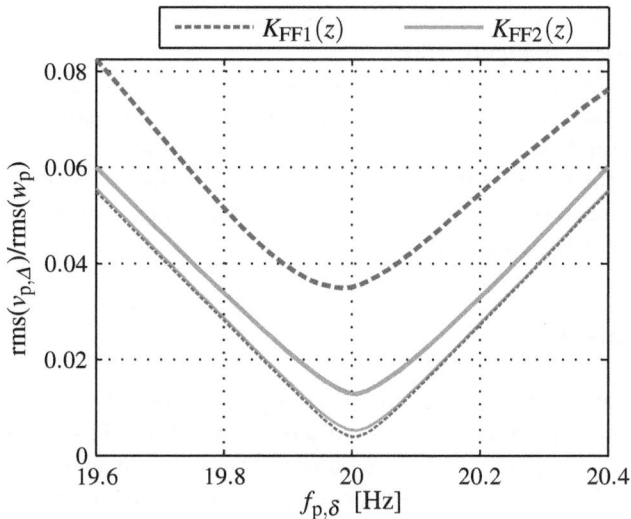

Fig. 3.10 Closed-loop rms reduction of the considered input $w_p(k)$ achieved by feedforward controllers $K_{FF1}(z)$ (optimal design) and $K_{FF2}(z)$ (optimal robust design) as a function of $f_{p,\delta}$: the thin and thick lines respectively correspond to the nominal and worst-case plant.

whereas Figure 3.9(b) evaluates their periodic performance for the worst-case plant:

$$V_{l,\text{wc}}/W_l = \max_{\omega \in \Omega_l} \left\{ |H_{p,\Delta}(\omega)|_{\text{wc}} \right\} .$$

Hence, Figures 3.9(a) and 3.9(b) are constructed from Figure 3.8(b) by computing the maximum of, respectively, the thin and thick curve over each of the gray-shaded uncertainty intervals.

Figure 3.8(b) reveals that $K_{FF1}(z)$ is very sensitive to plant uncertainty: at high frequencies where plant uncertainty is prominent, see Figure 3.5(b), $|H_p(\omega)|_{\text{wc}}$ deviates significantly from $|H_p(\omega)|$, both around and in between the harmonics. For $l \geq 13$ (260 Hz), $|H_{p,\Delta}(\omega)|_{\text{wc}} > 1$ in the gray-shaded uncertainty intervals, and hence, the periodic performance achieved by $K_{FF1}(z)$ for the worst-case plant is poor. This is clarified by Figure 3.9: at the higher harmonics, $V_{l,\text{wc}}$ is about 30 dB higher than the nominal value V_l.

Robust controller $K_{FF2}(z)$ is designed to yield good periodic performance for all potential plants $G_\Delta(z)$. Comparison of Equations 3.13 and 3.14 reveals that the difference between $|H_p(\omega)|$ and $|H_{p,\Delta}(\omega)|_{\text{wc}}$ can only be reduced by restricting the control action, that is: by reducing $|X(\omega)|$ and hence, $|K_{FF}(\omega)|$. This is confirmed by Figure 3.8(a), which shows that at high frequencies, $|X(\omega)|$ is significantly lower for $K_{FF2}(z)$ than for $K_{FF1}(z)$, particularly around the input harmonics. As revealed by Equation 3.13 and Figure 3.8(b), the reduced gain $|X(\omega)| \approx 0$ translates into $|H_p(\omega)| \approx 1$. Due to this property, at the higher harmonics, the periodic performance of $K_{FF2}(z)$ for the nominal plant is worse compared to $K_{FF1}(z)$, see Figure 3.9(a).

However, $K_{FF2}(z)$ yields significantly better periodic performance for the worst-case plant, as is clear from Figure 3.9(b).

Figure 3.10 evaluates for the two feedforward controllers the closed-loop performance for the particular reference trajectory $w_p(k)$ of Figure 3.4, by showing $\text{rms}(v_{p,\Delta}(k))/\text{rms}(w_p(k))$ as a function of $f_{p,\delta}$. The thin and thick lines respectively correspond to the nominal and worst-case plant. While for nominal plant $G(z)$, the overall performance of $K_{FF2}(z)$ is only slightly larger compared to $K_{FF1}(z)$, its worst-case performance is significantly better.

3.5 Conclusion

This chapter applies the general methodology of Chapter 2 to design a feedforward controller for periodic inputs to a discrete-time SISO LTI system. The design methodology is able to reproduce and outperform the design approach of Walgama and Sternby [152], currently the major feedforward controller design for periodic inputs. The latter design approach attributes all design freedom to eliminating the periodic input and hereby assumes perfect knowledge of input period and plant. The developed design methodology, on the other hand, provides the flexibility to incorporate additional design specifications and allows accounting for uncertainty on the input period as well as plant uncertainty. The latter two advantages are illustrated by numerical results.

Chapter 4
Application to Estimated Disturbance Feedback Control

4.1 Introduction

4.1.1 State of the Art

The appeal of feedforward control is to a large extent related to the following properties: (i) the only stability concern in a feedforward controller design is its own stability whereas a feedback controller must additionally guarantee stability of the closed-loop system; (ii) the effect of a feedforward controller is restricted to the input channel to which it is added, not affecting other inputs and the related performance; and (iii) many applications can cope with limited noncausality of a feedforward controller, whereas a feedback controller must be causal. Unfortunately, feedforward control only applies to reference inputs or measurable disturbances, while unmeasurable disturbances, on the other hand, are widespread in engineering practice. Extending the advantages of feedforward control to unmeasurable disturbances is the rationale for disturbance observers, where the observer developed by Ohnishi [103] is most popular [82, 95, 104, 116, 155]. Disturbance observers provide an estimate of the unmeasurable disturbance, which can be used to compute an appropriate control input similar to feedforward control.

In this monograph, the control strategy of feeding back an estimated disturbance to the control input is referred to as estimated disturbance feedback control[1], where the controller comprises a disturbance estimator and a disturbance feedback controller. Although the latter part features great similarity with a feedforward controller, its combination with a disturbance observer turns it into a feedback control strategy (hence its name). Hereby, estimated disturbance feedback control can only preserve the benefits of feedforward control to a limited extent: in the case of a perfect plant model, the estimated disturbance feedback controller does not affect closed-loop stability and transfer functions from alternative inputs, but these

---

[1] The literature reveals a variety of names for this control strategy, such as pseudo-feedforward control [151], virtual feedforward control [154], external model control [9, 127, 144, 158], etc. In addition, this control strategy is closely related to internal (plant) model control [102].

G. Pipeleers et al.: Optimal Linear Controller Design for Periodic Inputs, LNCIS 394, pp. 43–59.
springerlink.com   © Springer-Verlag Berlin Heidelberg 2009

benefits are compromised by model uncertainty [31, 32, 56, 130, 140, 153]. Moreover, in contrast to a feedforward controller, a disturbance feedback controller must be causal.

Due to the last property, for nonminimum-phase systems an estimated disturbance feedback controller design is even more challenging than a feedforward controller design [23, 129]. This motivated both Tomizika *et al.* [143] and Walgama and Sternby [152] to extend their feedforward controller design approach for periodic inputs to estimated disturbance feedback controllers. Tomizuka *et al.* [144, 158] apply a parameter adaptation algorithm to estimate the Fourier coefficients of the disturbance harmonics on line and according to [143] the control signal is computed to cancel the disturbance. Walgama [151] combines the FIR feedforward controller design of [152] with the disturbance observer of Ohnishi [103].

For persistent disturbances, which include periodic signals, disturbance observers are closely related to state observers, as their signal generator can be absorbed in the plant dynamics [70, 100, 114, 125]. In fact, translated to output regulation theory (see Appendix B for an introduction and [117] for an in-depth treatment), the design of [151] corresponds to the combination of a specific Luenberger state estimator [96] and a particular state-feedback controller. Relying on the equivalence between output and state observers, more advanced periodic disturbance estimators have been proposed in the literature [9, 100, 114, 127]. Their advantage is faster elimination of an initial estimation error, but at the downside they explicitly rely on the periodic signal generator (they only estimate the disturbance at the harmonic frequencies) and hereby cannot cope with period-time uncertainty. Therefore, these disturbance observers are not discussed in this chapter.

4.1.2   Contribution

This chapter deals with the design of an estimated disturbance feedback controller for periodic disturbances acting on a discrete-time SISO LTI system. The disturbance estimator of Ohnishi [103] is combined with a disturbance feedback controller, designed according to the methodology of Chapter 2. Hereby, the estimated disturbance feedback controller design of Walgama [151] is extended with the following advantages:

Period-time Uncertainty: Instead of enforcing perfect rejection of the periodic disturbance for the nominal period, the primal objective in the design methodology is minimizing the periodic performance index, which explicitly accounts for period-time uncertainty.

Multi-objective Control: Whereas in [151] all design freedom is attributed to eliminating the periodic disturbance, the developed design methodology allows accounting for a variety of additional design specifications, such as transient response, control effort, etc.

Plant Uncertainty: Contrary to [151], the design methodology allows accounting for plant uncertainty.

4.1.3   Outline

Section 4.2 provides some background on estimated disturbance feedback control
and reviews the disturbance feedback controller design of [151]. Section 4.3 applies
the developed methodology to design a disturbance feedback controller and elaborates on the corresponding general control configuration, Youla parametrization and
optimal design, while Section 4.4 illustrates its potential by numerical results.

4.2   Background

This section presents the architecture of an estimated disturbance feedback controller (Section 4.2.1) and briefly reviews the disturbance feedback controller design
of [151] (Section 4.2.2).

4.2.1   Control Configuration

Figure 4.1 shows, for a stable plant $G(z)$, the architecture of an estimated disturbance feedback controller that applies the disturbance observer of Ohnishi [103][2].
The disturbance observer computes the following estimate $\widehat{d}(k)$ of output disturbance $d(k)$ from the control signal $u(k)$ and plant output $\eta(k)$:

$$\widehat{d}(k) = \eta(k) - G(q)u(k) .$$

This estimate relies on the plant model $G(z)$ and if this model is accurate, the disturbance observer yields a perfect estimate: $\widehat{d}(k) = d(k)$. The second part of the
estimated disturbance feedback controller is the disturbance feedback controller
$K_{\mathrm{dFB}}(z)$ which feeds back $\widehat{d}(k)$ to the plant input.

  The estimated disturbance feedback controller can be applied to the open-loop
plant, Figure 4.1(a); or combined with a feedback controller $K_o(z)$, Figure 4.1(b).
In the latter case, $K_o(z)$ is assumed to be stable and designed *a priori*, yielding
the so-called original feedback system. This system must be internally stable and
its closed-loop sensitivity and complementary sensitivity are denoted by $S_o(z)$ and
$T_o(z)$, respectively:

---

[2] The derivation for an unstable plant proceeds along the same lines, but then an estimate of
$V_G(q)d(k)$ instead of $d(k)$ is computed similarly to Figure 4.1:

$$V_G(q)\widehat{d}(k) = V_G(q)\eta(k) - U_G(q)u(k) ,$$

·where stable transfer functions $U_G(z)$ and $V_G(z)$ constitute to a fractional representation of
$G(z)$: $G(z) = U_G(z)V_G(z)^{-1}$ [97].

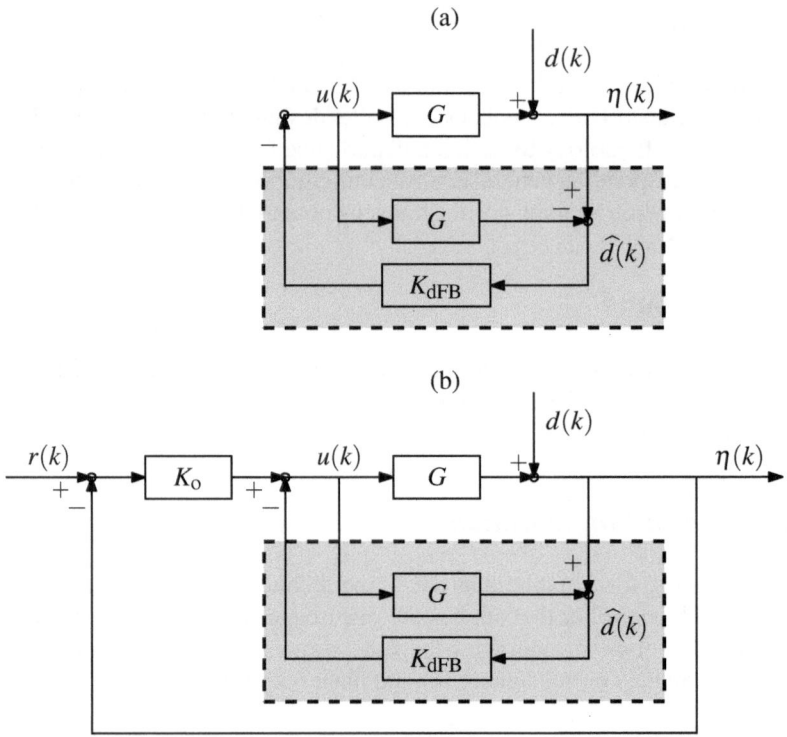

Fig. 4.1 Estimated disturbance feedback control: relying on plant model G, an estimate $\widehat{d}(k)$ of output disturbance $d(k)$ is constructed, which is fed back to the control signal $u(k)$ by the disturbance feedback controller K_{dFB}. The estimated disturbance feedback controller is either applied in open loop (a); or combined with a feedback controller K_o (b).

$$S_o(z) = \frac{1}{1 + K_o(z)G(z)}, \qquad T_o(z) = \frac{K_o(z)G(z)}{1 + K_o(z)G(z)}. \qquad (4.1)$$

As the control configuration of Figure 4.1(b) can reproduce Figure 4.1(a) by setting $K_o(z) = 0$, the remainder of this chapter is elaborated for the former setup.

Disturbance $d(k)$ is considered periodic and specified according to Section 2.2.2, yielding $w_p(k) = d(k)$ and $v_p(k) = \eta(k)$. In case of a perfect plant model, $K_{dFB}(z)$ acts as a feedforward controller: (i) $K_{dFB}(z)$ preserves the original closed-loop transfer functions from reference input $r(k)$; and (ii) as long as $K_{dFB}(z)$ is stable, it does not compromise closed-loop stability. The similarity with feedforward control also prevails in the closed-loop transfer function $H_p(z)$ from $w_p(k)$ to $v_p(k)$:

$$H_p(z) = S_o(z)\left[1 - G(z)K_{dFB}(z)\right], \qquad (4.2)$$

which resembles (3.1).

4.2.2 Current Design Approach

The estimated disturbance feedback controller design of Walgama [151] assumes both a perfect plant model and accurate knowledge of the input period. Since for a perfect plant model the disturbance feedback controller behaves like a feedforward controller, $K_{dFB}(z)$ is designed according to [152] (see Section 3.2.2). That is: it is set equal to a FIR filter of length n_Λ, Equation 2.6:

$$K_{dFB}(z) = \sum_{m=1}^{n_\Lambda} k_{dFB,m} z^{1-m}, \qquad (4.3)$$

where the filter coefficients $k_{dFB,m}$ are computed such that

$$H_p(l\omega_p) = S_o(l\omega_p)\left[1 - G(l\omega_p)K_{dFB}(l\omega_p)\right] = 0, \qquad \forall l \in \mathscr{L}.$$

Section 3.2.2 reviews four approaches to compute $k_{dFB,m}$ according to these constraints.

4.3 Application of the Design Methodology

This section applies the general methodology of Chapter 2 to design a disturbance feedback controller and elaborates on the corresponding general control configuration (Section 4.3.1), Youla parametrization (Section 4.3.2) and optimal design (Section 4.3.3). Section 4.3.4 renders the optimal disturbance feedback controller design robust for unstructured plant uncertainty[3].

4.3.1 General Control Configuration

Essentially, estimated disturbance feedback control constitutes a specific way of designing a general feedback controller $K(z)$ to attenuate an unmeasurable disturbance. Figure 4.2(a) shows the corresponding control configuration, which translates into the general control configuration shown in Figure 4.2(b). The corresponding generalized plant $P(z)$ is given by

$$\begin{bmatrix} v_p(k) \\ y(k) \end{bmatrix} = \underbrace{\begin{bmatrix} 1 & G(q) \\ -1 & -G(q) \end{bmatrix}}_{P(q)} \begin{bmatrix} w_p(k) \\ u(k) \end{bmatrix},$$

---

[3] Chapters 3 and 4 focus on rendering the developed design methodology robust for plant uncertainty, while Chapters 5 and 6 emphasize the methodology's multi-objective nature, trading off closed-loop periodic performance against conflicting design objectives.

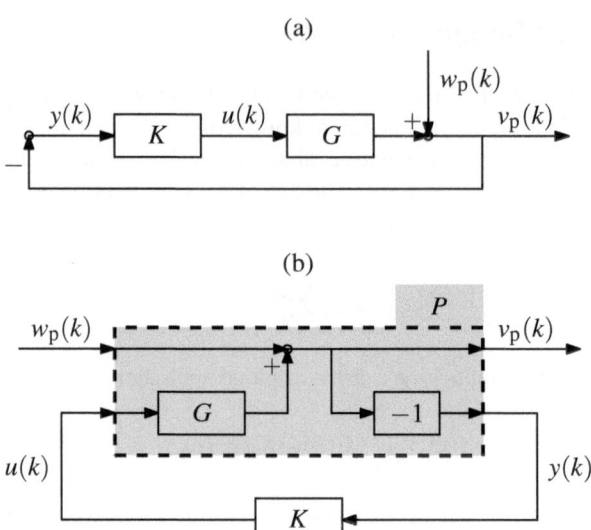

Fig. 4.2 General control configuration (b) for the design of feedback controller K to attenuate the periodic output disturbance $w_p(k)$ to the output $v_p(k)$ of plant G (a).

while accounting for additional design specifications requires extending $P(z)$ with complementary exogenous inputs and regulated outputs.

4.3.2   Youla Parametrization

The Youla parametrization augments the generalized plant with a nominal controller and hereby translates the controller design into the design of Youla parameter $X(z)$. The nominal controller $K_{nom}(z)$ is set equal to the original feedback controller $K_o(z)$, and the addition of the disturbance observer of Ohnishi [103] corresponds to a particular augmentation of $K_{nom}(z)$. This augmentation is shown in Figure 4.3 and yields

$$\begin{bmatrix} u(k) \\ \tilde{y}(k) \end{bmatrix} = \underbrace{\begin{bmatrix} K_o(q) & -1 \\ -(1+K_o(q)G(q)) & G(q) \end{bmatrix}}_{\widetilde{K}_{nom}(q)} \begin{bmatrix} y(k) \\ \tilde{u}(k) \end{bmatrix},$$

while the corresponding augmented plant $\widetilde{P}(z)$ is given by

$$\begin{bmatrix} v_p(k) \\ \tilde{y}(k) \end{bmatrix} = \underbrace{\begin{bmatrix} S_o(q) & -S_o(q)G(q) \\ 1 & 0 \end{bmatrix}}_{\widetilde{P}(q)} \begin{bmatrix} w_p(k) \\ \tilde{u}(k) \end{bmatrix}.$$

(a)

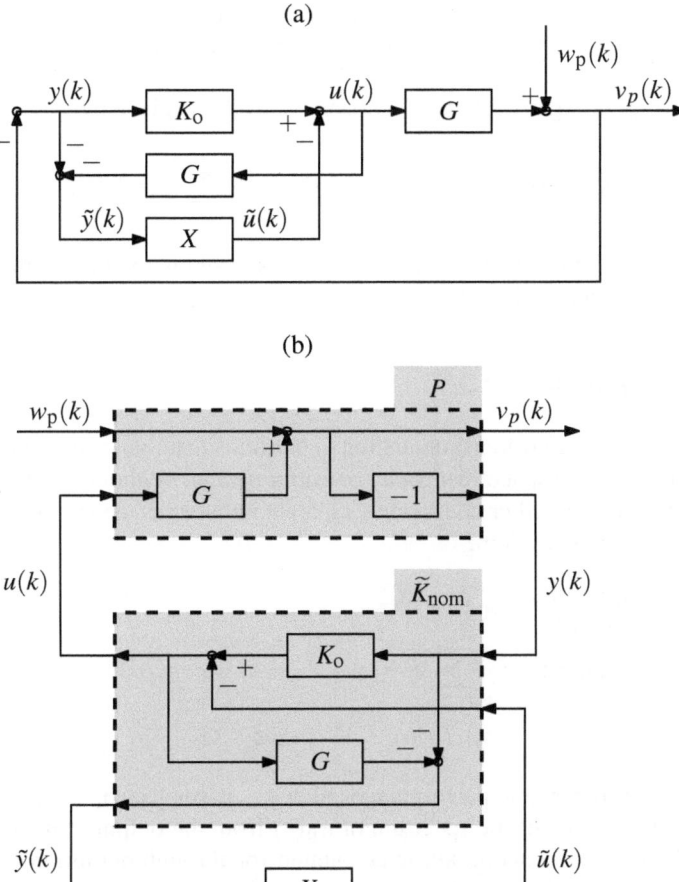

(b)

Fig. 4.3 Youla parametrization: augmentation of nominal controller $K_{\text{nom}} = K_{\text{o}}$ corresponding to the disturbance observer of Ohnishi [103]: (a) in the classical feedback control configuration; and (b) in the general control configuration.

Youla parameter $X(z)$ acts as a disturbance feedback controller, and application of parametrization (2.17) yields

$$K_{\text{dFB}}(z) = X(z) = \sum_{m=1}^{M} x_m z^{1-m} . \tag{4.4}$$

The design parameters x_m are grouped in the vector $x \in \mathbf{R}_M$, (2.16), and for $M \geq n_\Lambda$, this parametrization encompasses the disturbance feedback controller (4.3) of [151].

An alternative parametrization of the disturbance feedback controller is obtained by redefining $\tilde{u}(k)$ as $G_-(q)^{-1}\tilde{u}(k)$, where $G_-(z)$ denotes the invertible part of $G(z)$. This yields the following augmented controller:

$$\tilde{K}_{\text{nom}}(z) = \begin{bmatrix} K_o(z) & -G_-(z)^{-1} \\ -\left(1 + K_o(z)G(z)\right) & G_+(z) \end{bmatrix},$$

and the corresponding disturbance feedback controller is given by

$$K_{\text{dFB}}(z) = G_-(z)^{-1}X(z) = G_-(z)^{-1} \sum_{m=1}^{M} x_m z^{-m+1}. \tag{4.5}$$

Substituting relation (4.4) or (4.5) in Equation 4.2 yields the closed-loop transfer function $H_p(z)$ as a function of x.

4.3.3 Optimal Design

Good steady-state closed-loop attenuation of periodic disturbance $w_p(k)$ is the major objective in the disturbance feedback controller design. Without additional design specifications, the FIR filter coefficients x_m are computed as the solution of the following optimization problem:

$$\underset{x,\gamma_{p,2},V_l}{\text{minimize}} \quad \gamma_{p,2} \tag{4.6a}$$

$$\text{subject to} \quad \sqrt{\sum_{l \in \mathscr{L}} V_l^2} \leq \gamma_{p,2} \tag{4.6b}$$

$$W_l |H_p(\omega)| \leq V_l, \quad \forall \omega \in \Omega_l, \quad \forall l \in \mathscr{L}. \tag{4.6c}$$

The 2-norm based periodic performance index $\gamma_{p,2}$ is suggested, since the designer can obtain an estimate of the spectrum of $w_p(k)$ from the output of the disturbance observer. Period-time uncertainty is accounted for through definition (2.4) of the uncertainty intervals Ω_l around the harmonics.

Design problem (4.6) can be supplemented with additional design specifications, such as constraints on the transient response of $H_p(z)$, reduced actuator effort, limited effect of measurement noise, etc. (see e.g. [13] for an overview). In view of the discussion in Section 2.3.3, for a strictly causal plant $G(z)$ the disturbance feedback controller (which must be causal) is bound to a trade-off between periodic and non-periodic performance. That is: if the disturbance comprises both a periodic and a nonperiodic part, closed-loop attenuation of one part generally implies closed-loop amplification of the other part. The analysis of this trade-off is similar to the elaboration in Section 6.5.2, and therefore not further discussed here. Instead, it is illustrated below how to render optimization problem (4.6) robust for plant uncertainty.

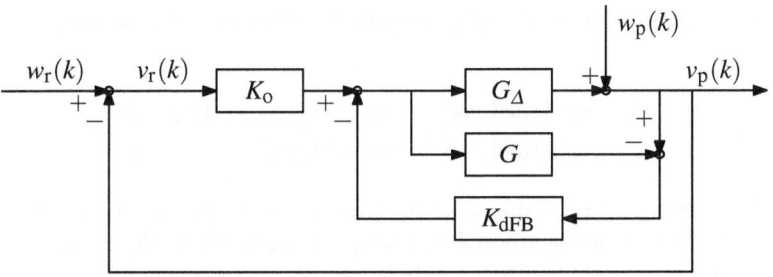

Fig. 4.4 Estimated disturbance feedback control configuration in the case of an uncertain plant G_Δ.

4.3.4   Optimal Robust Design for Plant Uncertainty

This section renders the optimal disturbance feedback controller design (4.6) robust for multiplicative unstructured plant uncertainty: all potential plant models $G_\Delta(z)$ are of the form (2.2). The disturbance observer incorporates the nominal plant model $G(z)$, and due to model uncertainty, disturbance estimate $\widehat{d}(k)$ differs from $d(k)$. This compromises the feedforward-like behavior of $K_{\mathrm{dFB}}(z)$ and, in contrast to its closed-loop effect for the nominal plant, $K_{\mathrm{dFB}}(z)$ does affect robust closed-loop stability and robust closed-loop performance for reference input $r(k)$.

For the robust disturbance feedback controller design, the control configuration of Figure 4.1(b) is translated into Figure 4.4. Besides attenuating the periodic output disturbance $w_{\mathrm{p}}(k) = d(k)$ to the plant output $v_{\mathrm{p}}(k) = \eta(k)$, the design of $K_{\mathrm{dFB}}(z)$ must also account for the robust closed-loop tracking performance. This design specification is labeled $i = i_{\mathrm{r}}$, and involves exogenous input $w_{\mathrm{r}}(k) = r(k)$ and regulated output $v_{\mathrm{r}}(k) = r(k) - \eta(k)$, where $(\cdot)_{\mathrm{r}}$ is shortened notation for $(\cdot)_{i_{\mathrm{r}}}$. The considered uncertain closed-loop transfer functions are given by

$$H_{\mathrm{p},\Delta}(z) = S_{\mathrm{o}}(z)\frac{1 - G(z)K_{\mathrm{dFB}}(z)}{1 + H_{\mathrm{stab}}(z)\Delta(z)} \, , \qquad \Delta(z) \in \boldsymbol{\Delta} \, , \qquad (4.7a)$$

$$H_{\mathrm{r},\Delta}(z) = S_{\mathrm{o}}(z)\frac{1 + K_{\mathrm{dFB}}(z)G(z)W_G(z)\Delta(z)}{1 + H_{\mathrm{stab}}(z)\Delta(z)} \, , \qquad \Delta(z) \in \boldsymbol{\Delta} \, , \qquad (4.7b)$$

where $\boldsymbol{\Delta}$ is given by (2.2b). Transfer function $S_{\mathrm{o}}(z)$ corresponds to the nominal closed-loop sensitivity function realized by $K_{\mathrm{o}}(z)$, see Equation 4.1, and

$$H_{\mathrm{stab}}(z) = \frac{\big[K_{\mathrm{dFB}}(z) + K_{\mathrm{o}}(z)\big]G(z)W_G(z)}{1 + G(z)K_{\mathrm{o}}(z)} \, . \qquad (4.8)$$

Rendering the design of $K_{\mathrm{dFB}}(z)$ robust for plant uncertainty $\boldsymbol{\Delta}$ requires three modifications to (4.6). First, robust stability must be added to the controller design as a hard constraint, since $\Delta(z)$ appears in the denominator of the closed-loop transfer functions (4.7) and hereby has the potential of making the closed-loop system unstable. By application of the Nyquist stability criterion, robust stability is equivalent to [131]

$$|1 + H_{\text{stab}}(\omega)\Delta(\omega)| \neq 0 , \quad \forall \omega \in [0, \pi f_{\text{s}}] , \; \forall \Delta(z) \in \boldsymbol{\Delta} , \tag{4.9a}$$

$$\Leftrightarrow |1 + H_{\text{stab}}(\omega)\Delta(\omega)| > 0 , \quad \forall \omega \in [0, \pi f_{\text{s}}] , \; \forall \Delta(z) \in \boldsymbol{\Delta} . \tag{4.9b}$$

At each frequency ω, the worst-case complex scalar $\Delta(\omega)$ (i.e., the one that minimizes the left-hand side) has modulus $|\Delta(\omega)| = 1$ and its phase aligns $[H_{\text{stab}}(\omega)\Delta(\omega)]$ along the negative real axis. Hence, robust stability is equivalent to

$$1 - |H_{\text{stab}}(\omega)| > 0 , \quad \forall \omega \in [0, \pi f_{\text{s}}] , \tag{4.9c}$$

$$\Leftrightarrow \|H_{\text{stab}}(z)\|_\infty < 1 . \tag{4.9d}$$

After the substitution of parametrization (4.4) or (4.5) for $K_{\text{dFB}}(z)$ in Equation 4.8, constraint (4.9d) corresponds to a convex constraint in the design parameters x, which complies with both gridding and the KYP lemma.

The second concern in the robust disturbance feedback controller design is robust periodic performance. To guarantee good periodic performance for all potential plant models $G_\Delta(z)$, for each harmonic $l \in \mathscr{L}$, constraint (4.6c) is replaced by

$$W_l |H_{\text{p},\Delta}(\omega)|_{\text{wc}} \leq V_{l,\text{wc}} , \quad \forall \omega \in \Omega_l , \tag{4.10a}$$

where

$$|H_{\text{p},\Delta}(\omega)|_{\text{wc}} = \max_{|\Delta| \leq 1} \left\{ \left| \frac{1 - GK_{\text{dFB}}}{1 + K_\text{o}G + (K_\text{o} + K_{\text{dFB}})GW_G\Delta} \right| \right\} , \tag{4.10b}$$

$$= \frac{|1 - GK_{\text{dFB}}|}{|1 + K_\text{o}G| - |(K_\text{o} + K_{\text{dFB}})GW_G|} . \tag{4.10c}$$

In the right-hand side, argument ω is omitted to save space. The transition from (4.10b) to (4.10c) follows from the worst-case uncertainty $\Delta(\omega)$ which has modulus $|\Delta(\omega)| = 1$ and its phase aligns $[(K_\text{o}(\omega) + K_{\text{dFB}}(\omega))G(\omega)W_G(\omega)\Delta(\omega)]$ opposite to $[1 + K_\text{o}(\omega)G(\omega)]$. After substitution of (4.4) or (4.5), the corresponding robust constraint

$$\frac{|1 - G(\omega)K_{\text{dFB}}(\omega)|}{|1 + K_\text{o}(\omega)G(\omega)| - |(K_\text{o}(\omega) + K_{\text{dFB}}(\omega))G(\omega)W_G(\omega)|} \leq V_{l,\text{wc}} , \quad \forall \omega \in \Omega_l \tag{4.10d}$$

is convex in x but, however, not simultaneously convex in x and $V_{l,\text{wc}}$. Hence, by convex optimization, one can only check whether a disturbance feedback controller (4.4) or (4.5) exists that satisfies (4.10d) for given $V_{l,\text{wc}}$. In addition, constraint (4.10d) only complies with the gridding solution approach.

Third, $K_{\text{dFB}}(z)$ affects the robust closed-loop performance related to reference input $w_\text{r}(k)$. Suppose that in the design of $K_\text{o}(z)$, good robust closed-loop tracking was quantified as $\|W_\text{r}(z)S_{\text{o},\Delta}(z)\|_\infty \leq 1$, for all $\Delta(z) \in \boldsymbol{\Delta}$, then preserving this robust tracking performance imposes the following constraint in the design of $K_{\text{dFB}}(z)$:

$$|W_r(\omega)||H_{r,\Delta}(\omega)|_{wc} \leq 1 , \quad \forall \omega \in [0, \pi f_s] . \tag{4.11a}$$

The computation of

$$|H_{r,\Delta}(\omega)|_{wc} = \max_{|\Delta| \leq 1} \left\{ \left| \frac{1 + K_{dFB}GW_G\Delta}{1 + K_oG + (K_o + K_{dFB})GW_G\Delta} \right| \right\} \tag{4.11b}$$

is not as intuitive as (4.10); it involves the structured singular value and is elaborated in Appendix C. Still, a convex constraint in x is obtained, which only complies with the gridding solution approach.

Adopting these three modifications to the optimal disturbance feedback controller design (4.6), yields the following optimization problem:

$$\underset{x}{\text{minimize}} \quad 0 \tag{4.12a}$$

$$\text{subject to} \quad \|H_{stab}(z)\|_\infty < 1 \tag{4.12b}$$

$$W_l |H_{p,\Delta}(\omega)|_{wc} \leq V_{l,wc} , \quad \forall \omega \in \Omega_l , \quad \forall l \in \mathscr{L} \tag{4.12c}$$

$$|W_r(\omega)||H_{r,\Delta}(\omega)|_{wc} \leq 1 , \quad \forall \omega \in [0, \pi f_s] . \tag{4.12d}$$

Scalars $V_{l,wc}$ are no longer considered as optimization variables. However, if these scalars all depend on one variable β: $V_{l,wc} = \beta \overline{V}_{l,wc}$, as for instance holds for definition (2.7) of γ_p, then complementing (4.12) with a bisection algorithm [15] provides an efficient way to minimize β.

4.4 Numerical Results

To illustrate the potential of the design methodology of Chapter 2 for an estimated disturbance feedback control problem, this section continues the simulation example of Section 3.4. As the major difference with Section 3.4, $w_p(k)$ is considered here as an unmeasurable output disturbance instead of a reference input. The estimated disturbance feedback controller design is applied in open loop, according to Figure 4.1(a), which corresponds to $S_o(z) = 1$. Youla parametrization (4.5) is adopted:

$$K_{dFB}(z) = G_-(z)^{-1}X(z) , \tag{4.13}$$

whereby the closed-loop transfer function $H_p(z)$, and hence, the design of x depends solely on the noninvertible part $G_+(z)$ of the plant $G(z)$:

$$H_p(z) = 1 - G(z)K_{dFB}(z) = 1 - G_+(z)X(z) . \tag{4.14}$$

If no plant uncertainty is considered, optimization problem (4.6) yields the same solution obtained in Section 3.4.2, which follows from the similarity between Equations 3.11, 3.12, and 4.13, 4.14. In the presence of plant uncertainty, on the other hand, the robust feedforward and disturbance feedback controller design generally yield different solutions. This is illustrated by Section 4.4.1, which shows that robust feedforward controller $K_{FF2}(z)$, designed in Section 3.4.3 is unacceptable as a disturbance feedback controller. Section 4.4.2 designs an appropriate robust disturbance feedback controller.

4.4.1 Feedforward Versus Disturbance Feedback Control

Figure 4.5 evaluates $K_{FF2}(z)$ as a disturbance feedback controller, where the shaded bands indicate the uncertainty intervals Ω_l (2.4) around the harmonics $l \in \mathcal{L}$. Figure 4.5(a) resumes the amplitude of its Youla parameter $X(z)$, shown in Figure 3.8(a). For the considered simulation example, robust stability requirement (4.9) imposes

$$|W_G(\omega)G(\omega)K_{dFB}(\omega)| < 1 , \quad \forall \omega \in [0, \pi f_s] , \qquad (4.15a)$$

$$\Leftrightarrow |X(\omega)| < \frac{1}{|W_G(\omega)G_+(\omega)|} , \quad \forall \omega \in [0, \pi f_s] , \qquad (4.15b)$$

where the right-hand side of (4.15b) is indicated by the thick dashed line in Figure 4.5(a). This figure reveals that $K_{FF2}(z)$ violates the robust stability requirement, where the violations of (4.15b) primarily occur in between the harmonics. Around the harmonics, $|X(\omega)|$ is already restricted by the robust feedforward controller design, in order to achieve good robust periodic performance.

Figure 4.5(b) shows the amplitude of the closed-loop system $H_{p,\Delta}(z)$ achieved by $K_{FF2}(z)$ as disturbance feedback controller: The thin line indicates the nominal amplitude $|H_p(\omega)|$:

$$|H_p(\omega)| = |1 - G_+(\omega)X(\omega)| ,$$

while the thick line corresponds to the worst-case amplitude $|H_{p,\Delta}(\omega)|_{wc}$:

$$|H_{p,\Delta}(\omega)|_{wc} = \frac{|1 - G_+(\omega)X(\omega)|}{1 - |W_G(\omega)G_+(\omega)X(\omega)|} . \qquad (4.16)$$

The thin curve corresponds to the result shown in Figure 3.8(b). Since in both the robust feedforward and disturbance feedback controller design, good periodic performance for the worst-case plant requires $|X(\omega)|$ to be small around the harmonics, $|H_{p,\Delta}(\omega)|_{wc}$ is acceptable around all harmonics, except $l = 13$ (260 Hz).

Although not clear from Figures 3.8(b) and 4.5(b), for $l = 0$, the robust performance of $K_{FF2}(z)$ is better as disturbance feedback controller than as feedforward controller. This is clarified by the comparison of (3.14) and (4.16): for $|H_p(0)| = 0$, the former formula yields $|H_{p,\Delta}(0)|_{wc} = |W_G(0)|$, while the latter yields $|H_{p,\Delta}(0)|_{wc} = 0$.

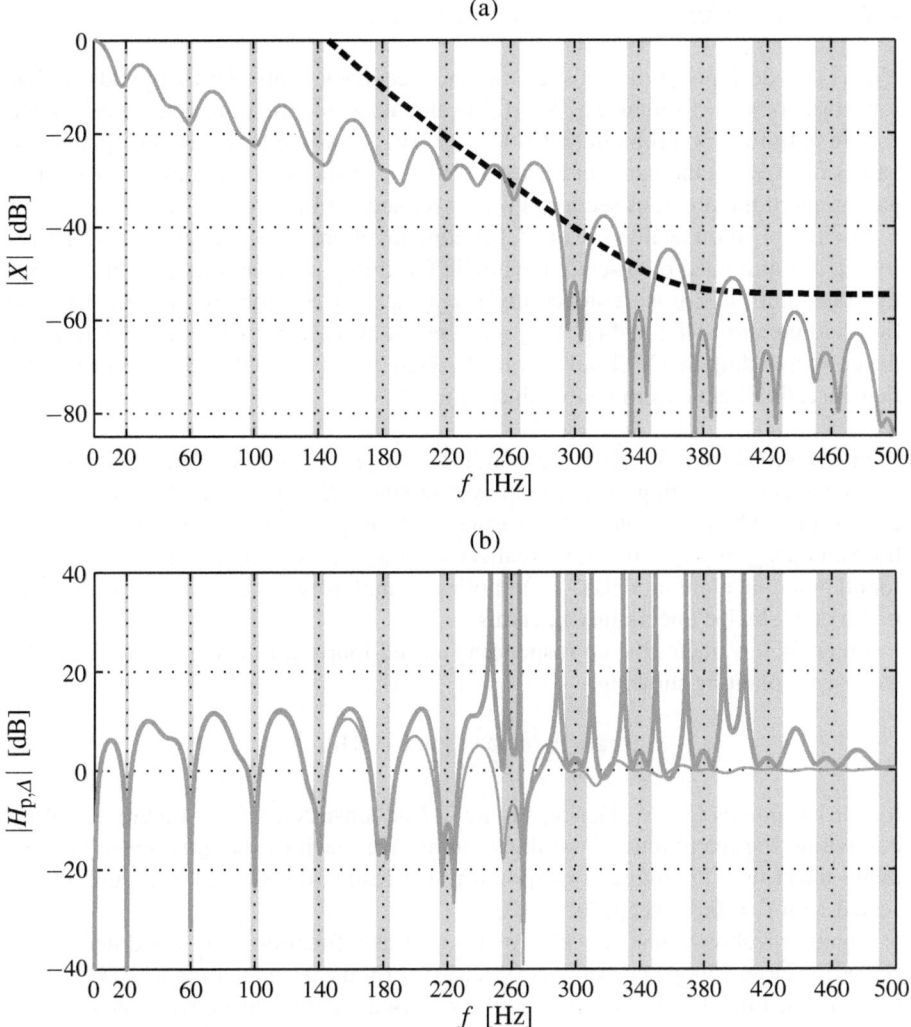

Fig. 4.5 Evaluation of feedforward controller $K_{FF2}(z)$ as a disturbance feedback controller: (a) amplitude FRF of Youla parameter $X(z)$, where robust stability requires $|X(\omega)|$ to reside below $1/|W_G(\omega)G_+(\omega)|$, indicated by the thick dashed line; and (b) amplitude FRF of the closed-loop transfer function $H_{p,\Delta}(z)$, where the thin and thick line respectively indicate the nominal amplitude $|H_p(\omega)|$ and worst-case amplitude $|H_{p,\Delta}(\omega)|_{wc}$.

4.4.2 Disturbance Feedback Controller

$K_{dFB1}(z)$ is designed as a robust disturbance feedback controller that yields similar robust periodic performance as $K_{FF2}(z)$ evaluated as feedforward controller. To that end, in optimization problem (4.12) $V_{l,wc} = \beta \overline{V}_{l,wc}$, where $\overline{V}_{l,wc}$ are set equal to the values for $K_{FF2}(z)$ indicated in Figure 3.9(b). According to the discussion in the last paragraph of the previous section, $\overline{V}_{0,wc}$ is reduced to $\overline{V}_{0,wc} = 10^{-6}$.

For $\beta = 1$, the feasible set of (4.12) contains more than one solution x. Therefore, β is minimized by bisection, while in feasibility problem (4.12), robust stability constraint (4.12b) is tightened to $\|H_{stab}(z)\|_\infty < 0.8$. Problem (4.12) is handled by gridding and SDPT3 [141, 149] requires on average 5.1 CPU seconds to solve the corresponding SOCP (Intel® Core™2 Duo T9300, 2.5 GHz, 3.5 GB of RAM). Few bisection iterates suffice to reduce β to 0.9.

Figure 4.6(a) shows $|X(\omega)|$ corresponding to $K_{dFB1}(z)$ and confirms robust stability of the corresponding closed-loop system since $|X(\omega)|$ resides below the thick dashed line, which indicates $1/|W_G(\omega)G_+(\omega)|$. Figure 4.6(b) shows $|H_p(\omega)|$ (thin line) and $|H_{p,\Delta}(\omega)|_{wc}$ (thick line) realized by $K_{dFB1}(z)$. Good periodic performance for all potential plant models $G_\Delta(z)$ implies a small difference between these curves in the gray-shaded uncertainty intervals.

Figure 4.7 evaluates the corresponding closed-loop periodic performance for the worst-case plant by showing

$$V_{l,wc}/W_l = \max_{\omega \in \Omega_l} \left\{ |H_{p,\Delta}(\omega)|_{wc} \right\} ,$$

for all harmonics $l \in \mathcal{L}$. Hence, Figure 4.7 is constructed from Figure 4.6(b) by computing the maximum of the thick curve, over each of the gray-shaded uncertainty intervals. This figure corresponds to the results for $K_{FF2}(z)$ in Figure 3.9(b), scaled with $\beta = 0.9$, except for $l = 0$.

Figure 4.8 shows $rms(v_{p,\Delta}(k))/rms(w_p(k))$ as a function of $f_{p,\delta}$ and hereby assesses the closed-loop performance of $K_{dFB1}(z)$ for the particular input $w_p(k)$ of Figure 3.4. The thick line corresponds to the worst-case plant and nearly coincides with the thin curve, which relates to the nominal plant. On account of $\beta = 0.9$ and the improved performance for $l = 0$, the worst-case reduction of $rms(w_p(k))$ achieved by $K_{dFB1}(z)$ is better than the result for $K_{FF2}(z)$ shown in Figure 3.10, for all $f_{p,\delta}$.

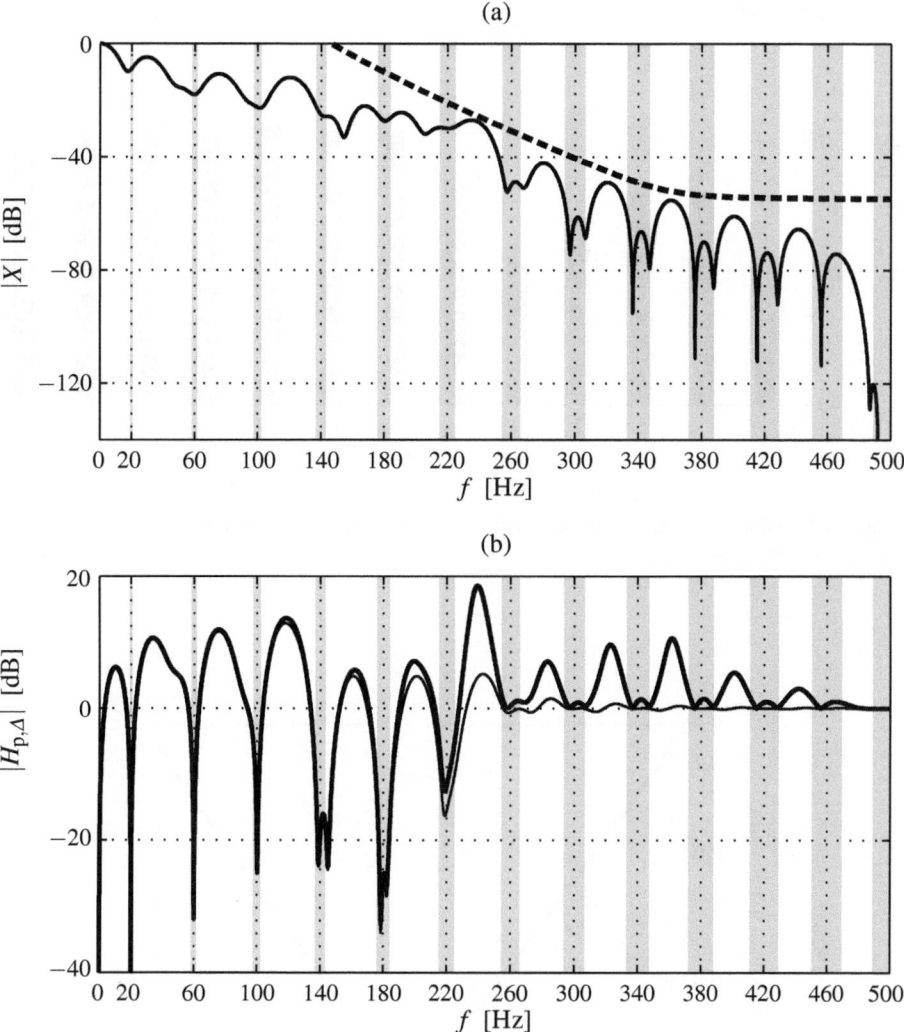

Fig. 4.6 Evaluation of optimal robust disturbance feedback controller $K_{\mathrm{dFB1}}(z)$: (a) amplitude FRF of Youla parameter $X(z)$, where robust stability requires $|X(\omega)|$ to reside below $1/|W_G(\omega)G_+(\omega)|$, indicated by the thick dashed line; and (b) amplitude FRF of the closed-loop transfer function $H_{\mathrm{p},\Delta}(z)$, where the thin and thick line respectively indicate the nominal amplitude $|H_{\mathrm{p}}(\omega)|$ and worst-case amplitude $|H_{\mathrm{p},\Delta}(\omega)|_{\mathrm{wc}}$.

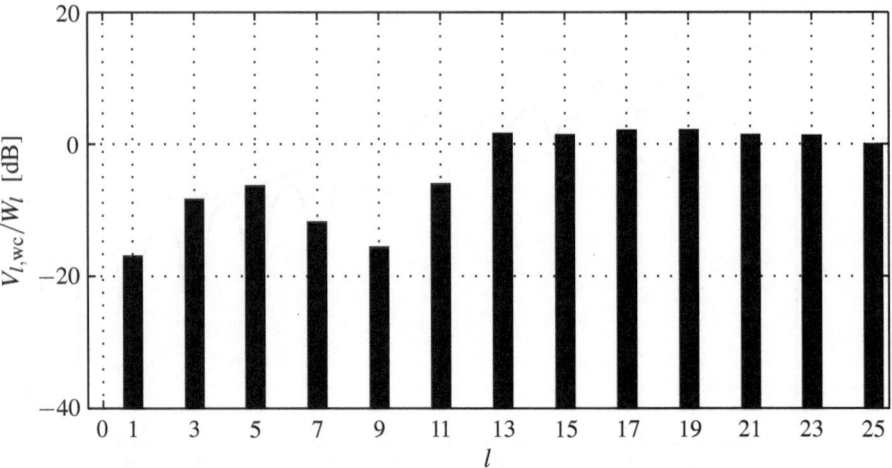

Fig. 4.7 Evaluation of optimal robust disturbance feedback controller $K_{\mathrm{dFB1}}(z)$: $V_{l,\mathrm{wc}}/W_l = \max_{\omega \in \Omega_l} \left\{ |H_{\mathrm{p},\Delta}(\omega)|_{\mathrm{wc}} \right\}$.

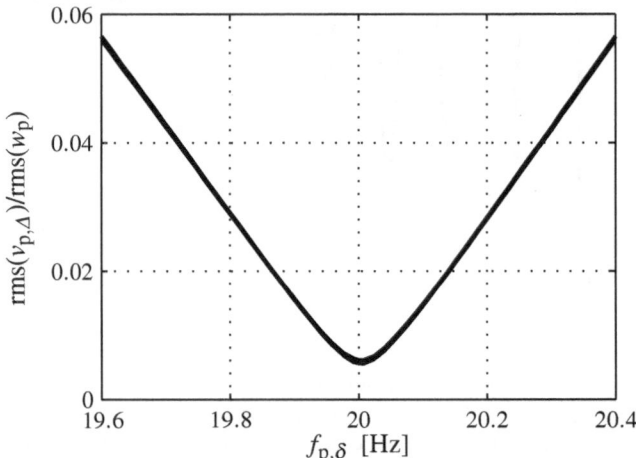

Fig. 4.8 Closed-loop rms reduction of the particular input $w_{\mathrm{p}}(k)$ of Figure 3.4 achieved by $K_{\mathrm{dFB1}}(z)$ as a function of $f_{\mathrm{p},\delta}$: the thick line corresponds to the worst-case plant and nearly coincides with the thin line, which relates to the nominal plant.

4.5 Conclusion

This chapter designs an estimated disturbance feedback controller for periodic disturbances acting on a discrete-time SISO LTI system. The controller comprises the most popular disturbance estimator of Ohnishi [103] and a disturbance feedback controller, designed according to the methodology of Chapter 2. Contrary to all design approaches from the literature, the proposed design methodology can cope with uncertainty on the input period. Moreover, the design methodology has the advantage of easy incorporation of plant uncertainty and additional design specifications. The potential of the resulting disturbance feedback controller design is illustrated by numerical results.

Chapter 5
Application to Repetitive Control

5.1 Introduction

5.1.1 State of the Art

In the early 1980s, Inoue *et al.* [75, 76] and Hara *et al.* [59, 60] laid the foundation of repetitive control. Facing the control problem of asymptotically rejecting periodic inputs of which only the period T_p is known, they relied on the Internal Model Principle [34, 45, 46, 47, 48] and included the signal generator shown in Figure 5.1(a) in the controller. This is the most general periodic signal generator as, determined by its initial conditions, it can generate any signal with period T_p. Although the first repetitive controllers were described in continuous-time, the discrete-time repetitive controller design got the upper hand [145, 146]. The preference for the discrete-time design stems from the digital controller implementation and the advantage that the discrete-time counterpart of the infinite-dimensional continuous-time signal generator of Figure 5.1(a) is of finite dimension, provided that T_p contains an integer number N of sample periods T_s. Figure 5.1(b) shows the corresponding discrete-time signal generator.

During the first decade, research on repetitive control focussed on closed-loop stability, as including the signal generator of Figures 5.1(a) and 5.1(b) is detrimental for stability [60]. To resolve the stability issues, the repetitive controller evolved from the signal generator of Figure 5.1(b) to the structure shown in Figure 5.1(c) [27, 28, 74, 128, 134]. Filter $L(z)$ guarantees nominal closed-loop stability by inverting the transfer function from $u_{RC}(k)$ to $y_{RC}(k)$ as determined by the control setup (see Section 5.2.2), while low-pass filter $Q(z)$ turns off the repetitive controller at high frequencies for reasons of robust stability [60, 61, 75]. Although more advanced repetitive controller designs have since been proposed [33, 55, 57, 66, 86, 91, 92, 105, 108, 132], the repetitive controller of Figure 5.1(c) remains most popular [25, 30, 40, 83, 99, 105, 132, 164].

G. Pipeleers et al.: Optimal Linear Controller Design for Periodic Inputs, LNCIS 394, pp. 61–82.
springerlink.com   © Springer-Verlag Berlin Heidelberg 2009

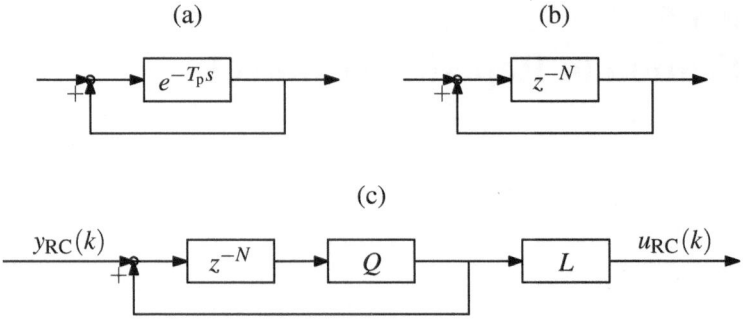

Fig. 5.1 Continuous-time (a) and discrete-time (b) generator of periodic signals with period $T_p = NT_s$; (c) corresponding structure of a discrete-time repetitive controller, where filters L and Q are added to ensure robust closed-loop stability.

By incorporating the signal generator of Figure 5.1(b), repetitive controllers yield the advantage of perfect nominal periodic performance: any periodic input is perfectly rejected/tracked asymptotically, provided that its period is exactly T_p. On the other hand, most repetitive controller designs from the literature suffer from two disadvantages. First, the obtained closed-loop periodic performance is very sensitive to uncertainty on the input period T_p. Hence, while yielding perfect nominal periodic performance, the resulting robust periodic performance is generally unsatisfactory. Consequently, well-functioning of these repetitive controllers requires the period T_p to be constant or accurately measurable, which may be jeopardized in practice by clock error drift, jitter, measurement noise, etc. Second, due to the Bode Integral Theorem [10, 21, 22, 49, 69, 138], the repetitive controller degrades the closed-loop performance for nonperiodic inputs, i.e., the nonperiodic performance: pushing the sensitivity down to zero at the multiples of ω_p is paid for by an increased sensitivity at intermediate frequencies.

To deal with these disadvantages, so-called high-order repetitive control has been proposed. Inoue [74] and Chang *et al.* [20] design high-order repetitive controllers to improve nonperiodic performance under the constraint of perfect nominal periodic performance, while the design of Steinbuch [135] improves robust periodic performance under the same constraint. A unified framework able to reproduce the results of both [20] and [135] is proposed in [136].

5.1.2 Contribution

This chapter applies the methodology of Chapter 2 to design a high-order repetitive controller for a discrete-time SISO LTI system. This results in a novel high-order repetitive controller design [112], which features the following advantages over the current design approaches [20, 74, 135, 136]:

Multi-objective Control: Contrary to all current design approaches, perfect periodic performance is not the starting-point of the presented repetitive controller design. Instead, the repetitive controller is designed to yield an optimal trade-off between performance indices γ_p and γ_{np}, which quantify its effect on the closed-loop periodic and nonperiodic performance, respectively.

Period-time Uncertainty: Periodic performance index γ_p explicitly accounts for period-time uncertainty, and this quantitative treatment of period-uncertain inputs contrasts the more qualitative approaches of [74] and [135].

Limits of Performance: The convex reformulation of the optimal design problem facilitates the computation of trade-off curves between conflicting performance indices γ_p and γ_{np}. By means of these trade-off curves the fundamental limits of performance in repetitive control are analyzed.

In addition, the proposed repetitive controller design approach is able to reproduce and outperform the designs by Chang *et al.* [20] and Steinbuch [135].

5.1.3   Outline

First, Section 5.2 details the control setup used in this chapter and presents the structure of a high-order repetitive controller. While this section also reviews the design approaches of [20] and [135], Section 5.3 applies the methodology of Chapter 2 to the high-order repetitive controller design. The potential of the resulting design approach is illustrated by numerical results presented in Section 5.4, while Section 5.5 summarizes the conclusions of this chapter.

5.2   Background

This section presents the control configuration of an add-on repetitive control system (Section 5.2.1) and briefly reviews the high-order repetitive controller designs of Chang *et al.* [20] and Steinbuch [135] (Section 5.2.2).

5.2.1   Control Configuration

As is common in the literature [25, 27, 40, 42, 81, 83, 135, 146], the repetitive controller, $K_{RC}(z)$, is considered as an add-on device, implying that it is added to the loop of an existing feedback system. This system is referred to as the "original feedback system" and comprises plant $G(z)$ and original feedback controller $K_o(z)$, which has been designed *a priori* and is hence considered fixed. Figure 5.2(a) shows the corresponding control setup, where signals $r(k)$, $d(k)$ and $\eta(k)$ respectively

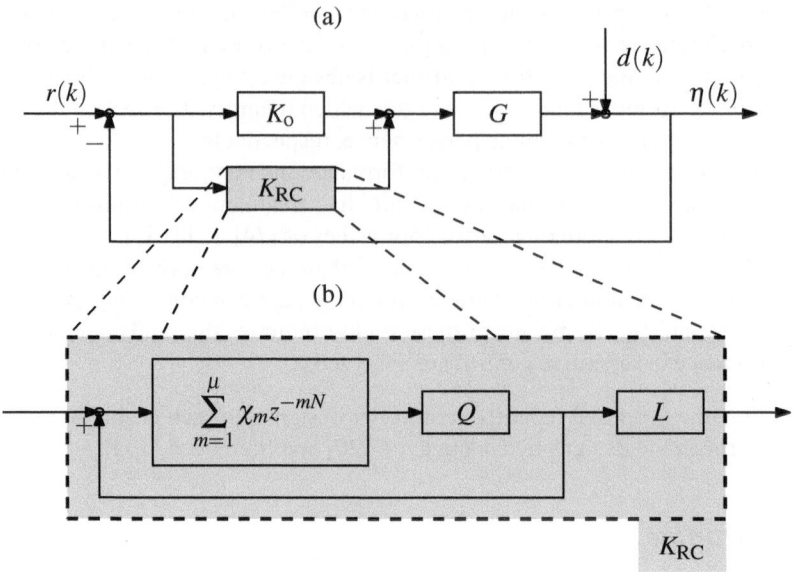

Fig. 5.2 (a) Add-on repetitive control configuration, where repetitive controller K_{RC} is added to the "original feedback system" comprising plant G and feedback controller K_o; and (b) structure of a μth-order repetitive controller. Signals $r(k)$, $d(k)$ and $\eta(k)$ respectively denote the reference trajectory, output disturbance and plant output.

correspond to the reference input, output disturbance and plant output. In closed loop, the tracking error $e(k) = r(k) - \eta(k)$ is given by

$$e(k) = S(q)\big(r(k) - d(k)\big) ,$$

where $S(q)$ corresponds to the closed-loop sensitivity. The combined input $r(k) - d(k)$ features both a periodic and a nonperiodic contribution:

$$r(k) - d(k) = w_p(k) + w_{np}(k) , \qquad (5.1)$$

where $w_p(k)$ is specified according to Section 2.2.2. The sample frequency is chosen such that T_p contains an integer number N of sample periods[1]:

$$T_p = N T_s .$$

---

[1] If N is not integer, it is rounded to the nearest integer N_{int} and the rounding error is accounted for as uncertainty on $T_p \equiv N_{int} T_s$.

Original Feedback System

$K_o(z)$ must yield an internally stable closed-loop system and is hence indispensable for unstable plants $G(z)$. The sensitivity and complementary sensitivity function of the original feedback system are respectively denoted by $S_o(z)$ and $T_o(z)$:

$$S_o(z) = \frac{1}{1 + K_o(z)G(z)}, \qquad T_o(z) = \frac{K_o(z)G(z)}{1 + K_o(z)G(z)}.$$

For ease of explanation, $K_o(z)$ is assumed to be stable and designed properly, which implies [131]: (i) a stable closed-loop system; (ii) high-gain feedback at low frequencies; (iii) sufficient roll-off of $|T_o(\omega)|$ at high frequencies; and (iv) a large modulus margin $\|S_o(z)\|_\infty^{-1}$. Figure 5.3 illustrates the FRFs of $S_o(z)$ and $T_o(z)$ corresponding to such a design.

Property (ii) is referred to as good nonperiodic performance, since it yields small $|S_o(\omega)|$ at low frequencies, resulting in a small tracking error $e(k)$ for any low-frequency input $w_{np}(k)$. By defining the bandwidth ω_{BW} of the original feedback system as the frequency where $|S_o(\omega)|$ first crosses $-3\,\mathrm{dB}$ from below, see Figure 5.3(a), $K_o(z)$ is said to yield good nonperiodic performance up to ω_{BW}.

Combination of properties (i), (iii) and (iv) ensures robust stability of the original feedback system. In the presence of multiplicative unstructured plant uncertainty (2.2), robust stability of the original feedback system requires $\|T_o(z)W_G(z)\|_\infty < 1$ [131], where optimizing closed-loop performance pushes $|T_o(\omega)|$ at high frequencies to its upper bound $|1/W_G(\omega)|$, as illustrated in Figure 5.3(b). The modulus margin $\|S_o(z)\|_\infty^{-1}$ corresponds to the minimal distance between the Nyquist plot of the loop transfer function and the point -1, and is therefore also considered as a robust stability measure.

Overall Feedback System

When $K_{RC}(z)$ is added to the loop, the closed-loop sensitivity changes from $S_o(z)$ to $S(z)$:

$$
\begin{aligned}
S(z) &= \frac{1}{1 + \left[K_o(z) + K_{RC}(z)\right]G(z)}, \\
&= S_o(z) \underbrace{\frac{1}{1 + K_{RC}(z)G(z)S_o(z)}}_{M_S(z)}.
\end{aligned}
$$

Transfer function $M_S(z)$ is called the modifying sensitivity function and represents the effect of $K_{RC}(z)$ on the closed-loop sensitivity. $K_{RC}(z)$ must not compromise the robust stability of the original feedback system, while improving closed-loop periodic performance. These design specifications for $K_{RC}(z)$ are nonconflicting provided that all harmonics $l \in \mathscr{L}$ lie well below ω_{BW}: at harmonics near or above ω_{BW}

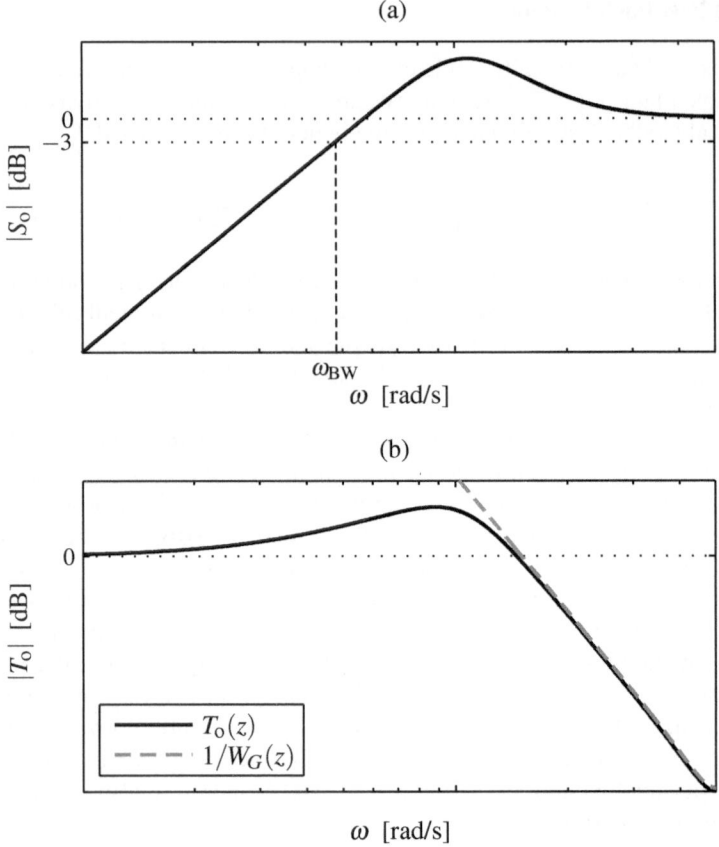

Fig. 5.3 Typical FRFs for the closed-loop sensitivity $S_{\mathrm{o}}(z)$ (a) and complementary sensitivity $T_{\mathrm{o}}(z)$ (b) of a properly designed original feedback system, where ω_{BW} indicates the bandwidth. In the presence of multiplicative unstructured plant uncertainty (2.2), robust closed-loop stability requires $\|T_{\mathrm{o}}(z)W_{G}(z)\|_{\infty} < 1$.

better performance simply cannot be achieved without compromising the modulus margin or high-frequency roll-off of the original feedback system. This assumption is made throughout this chapter.

High-Order Repetitive Controller

Figure 5.2(b) shows the structure of a repetitive controller of order μ. A high-order repetitive controller ($\mu > 1$) generalizes the classical, first-order repetitive controller shown in Figure 5.1(c) by extending z^{-N} to a polynomial in z^{-N}:

$$\chi(z) = \sum_{m=1}^{\mu} \chi_m z^{-mN} . \tag{5.2}$$

The repetitive controller structure of Figure 5.2(b) gives rise to the following expressions for $K_{RC}(z)$ and $M_S(z)$:

$$K_{RC}(z) = \frac{\chi(z)Q(z)L(z)}{1 - \chi(z)Q(z)} , \tag{5.3a}$$

$$M_S(z) = \frac{1 - \chi(z)Q(z)}{1 - \chi(z)Q(z)\left[1 - L(z)G(z)S_o(z)\right]} . \tag{5.3b}$$

5.2.2 Current Design Approaches

A high-order repetitive controller design involves designing $\chi(z)$ and the filters $Q(z)$ and $L(z)$. Current high-order repetitive controller designs [20, 74, 135, 136] adopt the design of $Q(z)$ and $L(z)$ that is common in first-order repetitive control [28] to guarantee preservation of the original feedback system's robust stability, independent of $\chi(z)$. In the second step, performance of the high-order repetitive controller is improved through the design of $\chi(z)$ [20, 74, 135, 136].

Design of $L(z)$: Nominal Stability

To achieve a nominally stable closed-loop system, $L(z)$ is set equal to

$$L(z) = \left[G(z)S_o(z)\right]^{-1} , \tag{5.4}$$

giving rise to a plant-independent expression of the modifying sensitivity function (5.3b):

$$M_S(z) = 1 - \chi(z)Q(z) . \tag{5.5}$$

However, design (5.4) poses two practical issues. First, nonminimum-phase zeros of $G(z)S_o(z)$ render $L(z)$ unstable, and hereby require the use of an approximate, stable inverse of $G(z)S_o(z)$. The "zero phase error tracking" inversion [142] is often used [61, 146], while alternative approximations are indicated in Section 3.1. In fact, by the effect of $Q(z)$ (discussed below), it suffices that $L(\omega) \approx [G(\omega)S_o(\omega)]^{-1}$ up to bandwidth ω_{BW} of the original feedback system.

The second practical issue raised by (5.4) is related to causality, since the inverse of a strictly causal $G(z)S_o(z)$, is noncausal. In addition, the approximate inversions of a nonminimum-phase system are generally noncausal. Fortunately, noncausality of $L(z)$ can be encompassed by absorbing it in the delay z^{-N}, as illustrated in Figure 5.4. Figure 5.4(b) shows the practical, causal implementation of the desired repetitive controller of Figure 5.4(a). The desired filter $L(z)$ is noncausal, while τ_L equals the smallest integer that renders $z^{-\tau_L}L(z)$ causal.

(a)

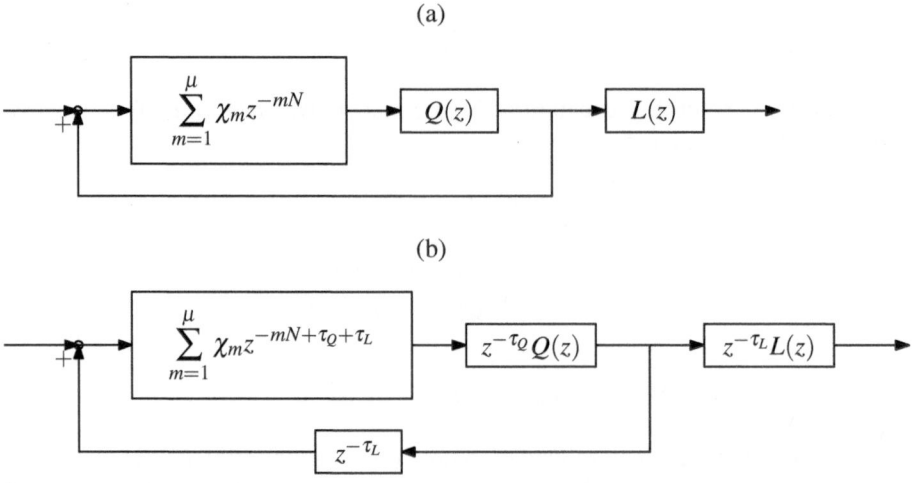

(b)

Fig. 5.4 Practical implementation (b) of repetitive controller (a) with noncausal filters $Q(z)$ and $L(z)$. Integers τ_L and τ_Q correspond to the smallest integers for which $z^{-\tau_L}L(z)$ and $z^{-\tau_Q}Q(z)$ are causal, and implementation (b) is causal provided that $N \geq \tau_Q + \tau_L$.

Design of $Q(z)$: Robust Stability

Robust closed-loop stability requires, in addition to nominal stability, preservation of the modulus margin and high-frequency roll-off of the original feedback system. To that end, the repetitive controller's action must be restricted to frequencies below the bandwidth ω_{BW} of the original feedback system. This is accomplished by designing $Q(z)$ as a low-pass filter where its order n_Q and cut-off frequency ω_Q are tuned such that all harmonics $l \in \mathscr{L}$ lie in the pass-band of $Q(z)$, while $|Q(\omega)| \approx 0$ for all $\omega \geq \omega_{BW}$ [28, 61, 136]. $Q(z)$ can comply with both requirements provided that all harmonics $l \in \mathscr{L}$ lie well below ω_{BW}, as is assumed here.

A zero-phase filter is preferred for $Q(z)$ and it is generally constructed from a linear-phase FIR filter $\widetilde{Q}(z)$. The phase behavior of $\widetilde{Q}(z)$ corresponds to a pure delay $z^{-\tau_Q}$, where $\tau_Q = n_Q/2$, such that

$$Q(z) = z^{\tau_Q}\widetilde{Q}(z)$$

has zero phase. However, $Q(z)$ is noncausal, while $\widetilde{Q}(z) = z^{-\tau_Q}Q(z)$ is causal. As illustrated in Figure 5.4, noncausality of $Q(z)$ can be compensated for by absorbing z^{τ_Q} in z^{-N}. The resulting practical implementation of the repetitive controller is causal provided that $N \geq \tau_Q + \tau_L$.

To illustrate the effect of $Q(z)$ on $M_S(z)$, Figure 5.5 compares, for a common repetitive control system, the FRF of $M_S(z)$ with the FRF of $\overline{M}_S(z)$, defined as the modifying sensitivity function for $Q(z) \equiv 1$:

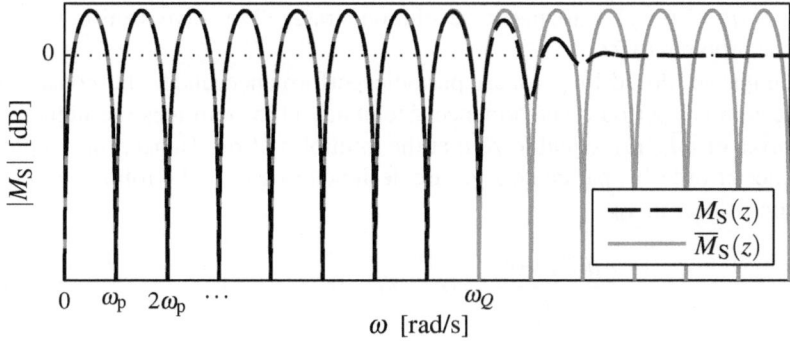

Fig. 5.5 Comparison between $M_S(\omega)$ and $\overline{M}_S(\omega)$, defined as the modifying sensitivity function for $Q(z) \equiv 1$, for a common repetitive control system. Up to cut-off frequency ω_Q of $Q(z)$, the curves coincide.

$$\overline{M}_S(z) = 1 - \chi(z) . \tag{5.6}$$

Since $\overline{M}_S(z)$ only contains powers of z^{-N}, its FRF is periodic with ω_p. Low-pass filter $Q(z)$ turns off $K_{RC}(z)$ from its cut-off frequency ω_Q, since $Q(\omega) \approx 0$ yields $M_S(\omega) \approx 1$. All harmonics $l \in \mathscr{L}$ are situated to the left of ω_Q, while $\omega_{BW} > \omega_Q$.

Design of $\chi(z)$: Performance

All current high-order repetitive controller designs adopt the assumption of a properly designed filter $Q(z)$, such that

$$M_S(l\omega_p) \approx \overline{M}_S(l\omega_p) , \quad \forall l \in \mathscr{L} .$$

This assumption justifies the design of $\chi(z)$ based on $\overline{M}_S(z)$ (5.6) instead of $M_S(z)$ (5.5).

Chang *et al.* [20] use χ_m to improve closed-loop nonperiodic performance under the constraint of perfect nominal periodic performance. To that end, parameters χ_m are computed as the solution of the following optimization problem:

$$\underset{\chi_m}{\text{minimize}} \quad \|\overline{M}_S(z)\|_\infty \tag{5.7a}$$

$$\text{subject to} \quad \sum_{m=1}^{\mu} \chi_m = 1 \tag{5.7b}$$

$$0 \le \chi_m \le 1 , \quad \forall m = 1, \ldots, \mu . \tag{5.7c}$$

Constraint (5.7b) guarantees perfect nominal periodic performance since it yields $\overline{M}_S(0) = 0$ and, implied by the FRF periodicity, $\overline{M}_S(l\omega_p) = 0$ for all integers l.

Chang *et al.* [20] use a stochastic "evolution strategy" to solve semi-infinite optimization problem (5.7).

To improve closed-loop robust periodic performance under the constraint of perfect nominal periodic performance, Steinbuch [135] enforces the higher-order derivatives of $|\overline{M}_S(\omega)|$ equal to zero at the multiples of ω_p. Hence, for a repetitive controller of order μ, parameters χ_m are designed to satisfy the following set of μ equations, linear in χ_m:

$$\frac{d^i |\overline{M}_S(0)|}{d\,\omega^i} = 0, \quad \forall i = 0, \ldots, \mu - 1. \tag{5.8}$$

5.3 Application of the Design Methodology

This section applies the methodology of Chapter 2 to design a high-order repetitive controller. Through parameters χ_m, the controller is designed to yield an optimal trade-off between performance indices γ_{np} and γ_p, which quantify its effect on the closed-loop nonperiodic and robust periodic performance. The design of filters $Q(z)$ and $L(z)$ is preserved from Section 5.2.2.

This section directly addresses the optimal design of χ_m, while the general control configuration and Youla parametrization are temporarily omitted. They are picked-up again in Chapter 6, where a high-order repetitive controller is shown to correspond to a feedback controller with restricted design freedom in the Youla parameter.

5.3.1 Optimal Design

As presented in Section 5.2.1, the combined input $r(k) - d(k)$ features both a periodic and a nonperiodic contribution (5.1). To focus on the effect of the repetitive controller on the closed-loop performance, its performance is related to the modifying sensitivity function $M_S(z)$ instead of $S(z)$, yielding

$$v_{np}(k) = v_p(k) = \frac{1}{S_o(q)} e(k), \tag{5.9a}$$

$$H_{np}(z) = H_p(z) = \frac{S(z)}{S_o(z)} = M_S(z). \tag{5.9b}$$

The effect of the repetitive controller on the closed-loop nonperiodic performance is quantified by index γ_{np} (2.10):

$$\gamma_{np} = \|M_S(z)\|_\infty, \tag{5.10a}$$

$$\approx \|\overline{M}_S(z)\|_\infty, \tag{5.10b}$$

where the close relation between $\|M_S(z)\|_\infty$ and $\|\overline{M}_S(z)\|_\infty$ is clarified by Figure 5.5.

Applying ∞-norm based definition (2.7) of the periodic performance index yields

$$\gamma_{\mathrm{p}} = \max_{l \in \mathscr{L}} \left\{ W_l \max_{\omega \in \Omega_l} \{|M_{\mathrm{S}}(\omega)|\} \right\} , \tag{5.11a}$$

where, by the use of (5.9), it is appropriate to change weights W_l to $W_l|S_{\mathrm{o}}(l\omega_{\mathrm{p}})|$. The assumption that all harmonics $l \in \mathscr{L}$ lie in the pass-band of $Q(z)$, justifies the following approximation of γ_{p}:

$$\gamma_{\mathrm{p}} \approx \max_{l \in \mathscr{L}} \left\{ W_l \max_{\omega \in \Omega_l} \{|\overline{M}_{\mathrm{S}}(\omega)|\} \right\} , \tag{5.11b}$$

$$= \max_{l \in \mathscr{L}} \left\{ W_l \max_{\omega \in \overline{\Omega}_l} \{|\overline{M}_{\mathrm{S}}(\omega)|\} \right\} , \tag{5.11c}$$

where $\overline{\Omega}_l$ corresponds to Ω_l, (2.4), shifted around the origin:

$$\overline{\Omega}_l = \left[-l\omega_{\mathrm{p}}\delta , \, l\omega_{\mathrm{p}}\delta \right] . \tag{5.12}$$

Transition from (5.11b) to (5.11c) relies on the FRF periodicity of $\overline{M}_{\mathrm{S}}(z)$.

In a μth-order repetitive controller design, indices γ_{p} and γ_{np} are conflicting, since the Bode Integral Theorem [10, 21, 22, 49, 69, 138] dictates

$$\int_0^{\omega_{\mathrm{p}}} \log \left(|\overline{M}_{\mathrm{S}}(\omega)| \right) d\omega = 0 . \tag{5.13}$$

Hence, improved periodic performance, $\gamma_{\mathrm{p}} < 1$, comes at the price of nonperiodic performance degradation, $\gamma_{\mathrm{np}} > 1$. This trade-off between γ_{p} and γ_{np} is analyzed by solving the following optimization problem for various weights $\alpha \geq 0$:

$$\underset{\chi_m, \gamma_{\mathrm{p}}, \gamma_{\mathrm{np}}}{\text{minimize}} \quad \gamma_{\mathrm{p}} + \alpha \gamma_{\mathrm{np}} \tag{5.14a}$$

$$\text{subject to} \quad \|\overline{M}_{\mathrm{S}}(z)\|_\infty \leq \gamma_{\mathrm{np}} \tag{5.14b}$$

$$W_l |\overline{M}_{\mathrm{S}}(\omega)| \leq \gamma_{\mathrm{p}} , \quad \forall \omega \in \overline{\Omega}_l , \quad \forall l \in \mathscr{L} . \tag{5.14c}$$

Approximate definitions (5.10b) and (5.11b) are used since: (i) they are used in all current high-order repetitive controller designs; (ii) they are generally very accurate; and (iii) the resulting optimization problem is solved much more efficiently compared to the program formulated in terms of $M_{\mathrm{S}}(z)$. The latter advantage is related to the fact that $\overline{M}_{\mathrm{S}}(z)$ only comprises powers of z^{-N}. Hereby, $\overline{M}_{\mathrm{S}}(z)$ is equivalently described by a μth-order FIR filter at sample period $NT_{\mathrm{s}} = T_{\mathrm{p}}$, while at the actual sample period T_{s}, it corresponds to a FIR filter of order μN. Accordingly, application of the (generalized) KYP lemma yields LMIs involving matrix variables in $\mathbf{S}_\mu$ instead of $\mathbf{S}_{\mu N}$, which yields a significant reduction in computational time. Also the gridding solution approach benefits from the use of $\overline{M}_{\mathrm{S}}(z)$ instead of $M_{\mathrm{S}}(z)$, as the

FRF periodicity of $\overline{M}_S(z)$ allows reducing the frequency range involved in (5.14b) from $[0, f_s/2]$ to $[0, f_p/2]$.

Optimization problem (5.14) does not depend on the original feedback system to which the repetitive controller is added, and by the FRF periodicity of $\overline{M}_S(z)$, the solution of (5.14) is independent of ω_p. If all harmonics $l \in \mathcal{L}$ are accounted for with equal weights W_l, the set of constraints (5.14c) is equivalent to

$$|\overline{M}_S(\omega)| \leq \gamma_p , \quad \forall \omega \in \overline{\Omega}_{l_{max}} ,$$

where $l_{max} = \max_{l \in \mathcal{L}}\{l\}$, since for all $l \in \mathcal{L}$: $\overline{\Omega}_l \subset \overline{\Omega}_{l_{max}}$. In this special case, the solution of (5.14) only depends on $\mathcal{L}$ and δ through the product $l_{max}\delta$.

5.4 Numerical Results

This section illustrates the potential of the proposed high-order repetitive controller design by numerical results. The convexity of optimal design problem (5.14) facilitates the computation of trade-off curves between γ_p and γ_{np} for a fixed order μ, and these trade-off curves are discussed in Section 5.4.1. Section 5.4.2 analyzes how the trade-off between γ_p and γ_{np} evolves as a function of μ, and investigates the solution of (5.14) for $\mu \to \infty$. Subsequently, Section 5.4.3 compares the proposed high-order repetitive controller design with the current design approaches of [20] and [135].

In this section, equal weights $W_l = 1$ are applied to all $l \in \mathcal{L}$ and this way, the presented results only depend on $\mathcal{L}$ and δ through the product $l_{max}\delta$, for which various values are considered. In addition, the results are independent of the original feedback system to which the repetitive controller is added, and do not depend on ω_p either.

5.4.1 Trade-off $\gamma_p - \gamma_{np}$ for Fixed μ

Figure 5.6 shows the trade-off curves between γ_p and γ_{np}, for repetitive controllers of order $\mu = 1, 2, \ldots, 5$, while relative uncertainty $l_{max}\delta$ ranges from 0% in Figure 5.6(a), to 10% in Figure 5.6(d). These trade-off curves are computed by solving (5.14) for various weights $\alpha \geq 0$: by increasing α they are traced from left to right. Optimization problem (5.14) is rendered numerically tractable by application of the (generalized) KYP lemma and the resulting SDP is solved with the solver of Liu and Vandenberghe [94]. Each of the problems (5.14) required to generate Figure 5.6 is solved within 0.1 CPU second (Intel® Core™2 Duo T9300, 2.5 GHz, 3.5 GB of RAM).

The trade-off curves shown in Figure 5.6 indicate the limits of performance for μth-order repetitive controllers. For a given level of robust periodic performance improvement γ_p, the curve indicates the minimal level of nonperiodic performance

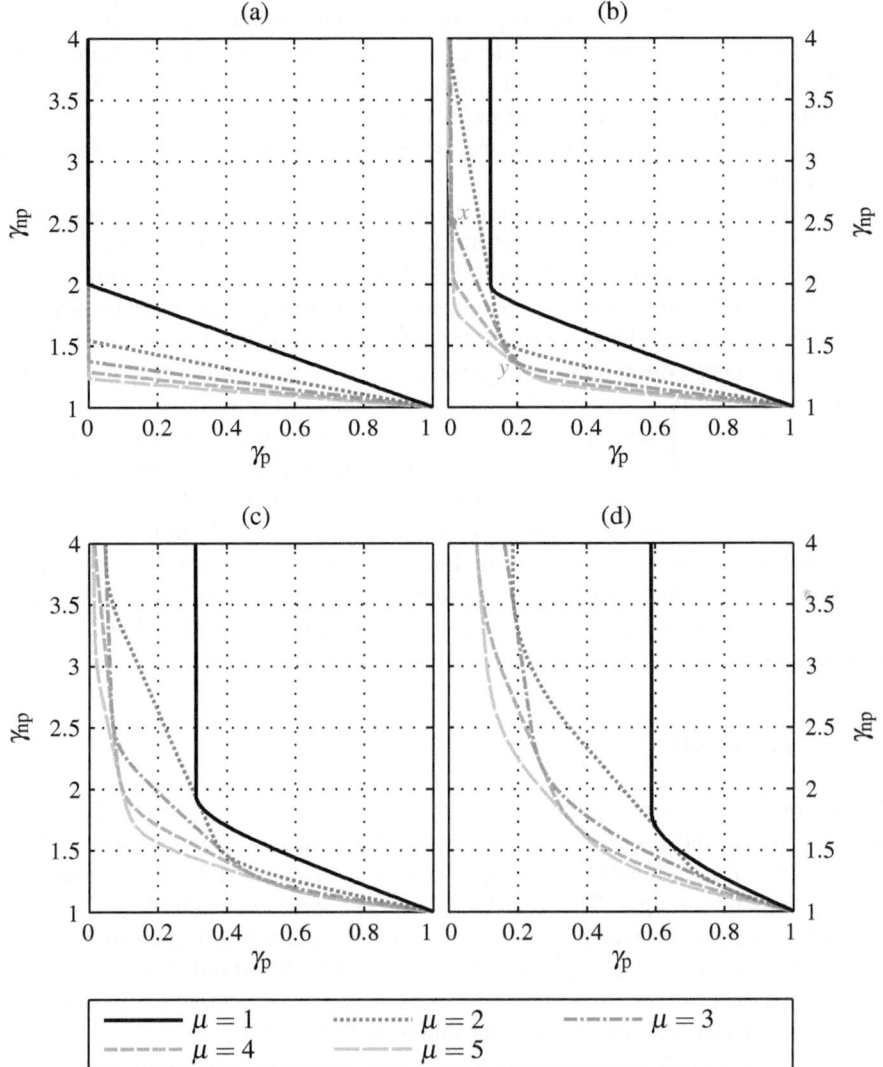

Fig. 5.6 Trade-off curves between γ_p and γ_{np} for different orders μ of the repetitive controller and various uncertainty levels $l_{max}\delta$ on the fundamental frequency: (a) $l_{max}\delta = 0\%$; (b) $l_{max}\delta = 2\%$; (c) $l_{max}\delta = 5\%$; and (d) $l_{max}\delta = 10\%$.

degradation γ_{np} that has to be tolerated. Or, *vice versa*, for a fixed level of allowable nonperiodic performance degradation, the trade-off curve indicates the best robust periodic performance improvement that a μth-order repetitive controller can achieve.

Given the higher number of design variables, a repetitive controller of a given order μ yields at least as good a performance as lower-order controllers, for all values of $l_{max}\boldsymbol{\delta}$: the trade-off curves shift to the left and/or downwards as μ increases. However, increasing the order does not guarantee better performance: Figures 5.6(b), 5.6(c) and 5.6(d) reveal points at which the trade-off curves for orders μ and $\mu + 1$ touch. In such a point neither γ_p nor γ_{np} improves if the order of the repetitive controller is increased by one, whereas increasing the order by two again results in improved performance.

Every trade-off curve is defined for γ_p values between an uncertainty and order-dependent lower limit and an uncertainty and order-independent upper limit $\gamma_p = 1$, corresponding to $K_{RC}(z) = 0$. If no uncertainty is present, the lower limit equals $\gamma_p = 0$ for all orders μ, whereas for a nonzero uncertainty, $\gamma_p = 0$ cannot be achieved by a finite-order controller since it invokes infinitely many equality constraints. The latter fact is clearly observable in Figure 5.6(d), whereas Figure 5.6(b) reveals that for small uncertainty levels, nearly zero γ_p values are still achievable.

The presence of "knees", i.e., abrupt changes in the slope of the trade-off curve, such as the points x and y for $\mu = 3$ in Figure 5.6(b), facilitates the choice of a good engineering design: y would be preferred over any repetitive controller design with higher γ_p, since the cost for decreasing γ_p to 0.18 is only small an increase in γ_{np}. On the other hand, is it not advisable to decrease γ_p below $1.23 \cdot 10^{-2}$ (point x) since this implies a very large increase in γ_{np}. If no uncertainty, Figure 5.6(a), or larger relative uncertainty levels, Figure 5.6(d), are considered, such clear knees are not present, whereby many more viable designs emerge.

5.4.2   Trade-off $\gamma_p - \gamma_{np} - \mu$

As the trade-off between γ_p and γ_{np} improves as μ increases, a trade-off surface between γ_p, γ_{np} and μ emerges. μ determines the transient response time of $S(z)$, since $M_S(z)$ has a finite impulse response of length $N\mu + n_Q/2$. Hence, in repetitive control, the Bode Integral Theorem (5.13) essentially dictates a trade-off between closed-loop periodic performance, nonperiodic performance and transient response time, while for a μth-order repetitive controller this trade-off reduces to a trade-off between periodic and nonperiodic performance. This section analyzes the $\gamma_{np} - \gamma_p - \mu$ trade-off surface and investigates the solution of (5.14) for $\mu \to \infty$.

Trade-off Surface

To analyze the trade-off between closed-loop periodic performance, nonperiodic performance and transient response time in repetitive control, Figure 5.7 shows for various uncertainty levels $l_{max}\boldsymbol{\delta}$, the trade-off curves between γ_p and γ_{np} for μ ranging from 1 to 100. These curves define the $\gamma_p - \gamma_{np} - \mu$ trade-off surface, which features a staircase-like behavior in μ, since μ is integer. The trade-off curves mark the μ stairs, where a repetitive controller of order μ can obtain all $\gamma_p - \gamma_{np}$ pairs above the corresponding trade-off curve.

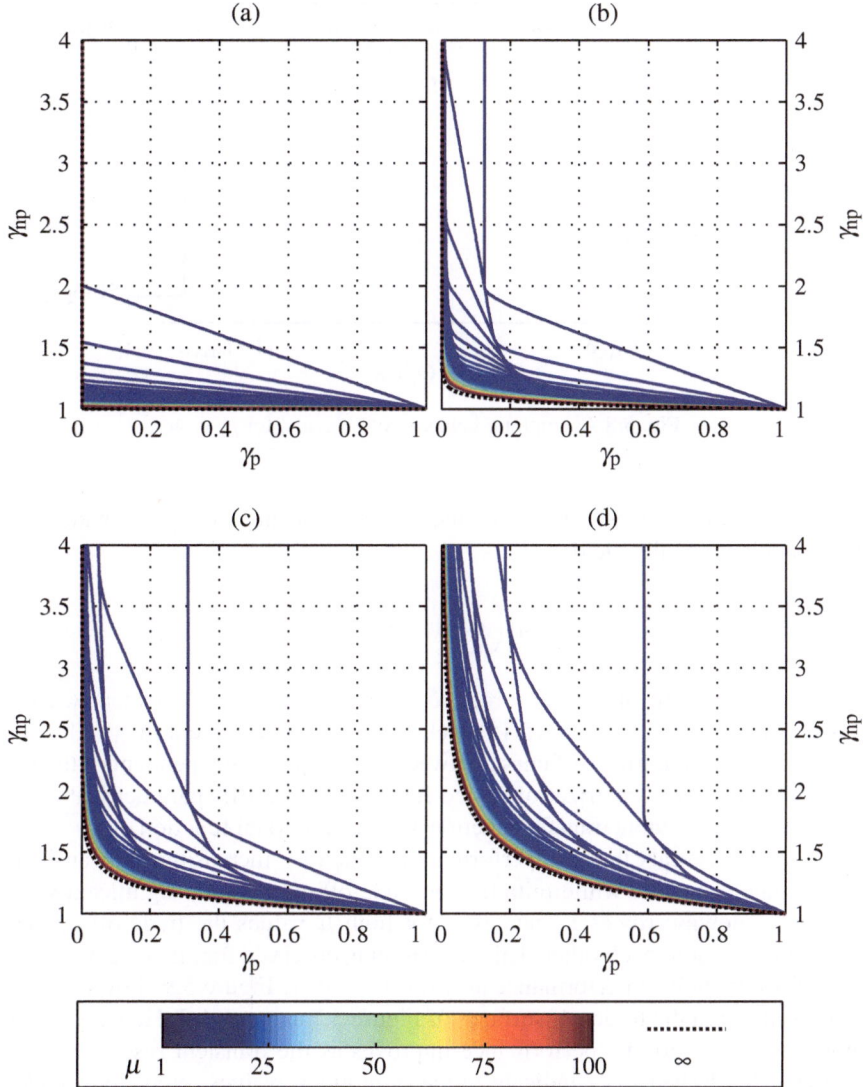

Fig. 5.7 Trade-off curves between γ_p and γ_{np} for μ ranging from 1 to 100, while the black dotted line corresponds to the asymptotic curve for $\mu \to \infty$. Four uncertainty levels $l_{max}\delta$ on the fundamental frequency are considered: (a) $l_{max}\delta = 0\%$; (b) $l_{max}\delta = 2\%$; (c) $l_{max}\delta = 5\%$; and (d) $l_{max}\delta = 10\%$.

The black dotted line in Figure 5.7 corresponds to the asymptotic trade-off curve between γ_p and γ_{np} for $\mu \to \infty$, and this curve is computed based on the asymptotic behavior of $|\overline{M}_S(\omega)|$ for $\mu \to \infty$, shown in Figure 5.8. The Bode Integral

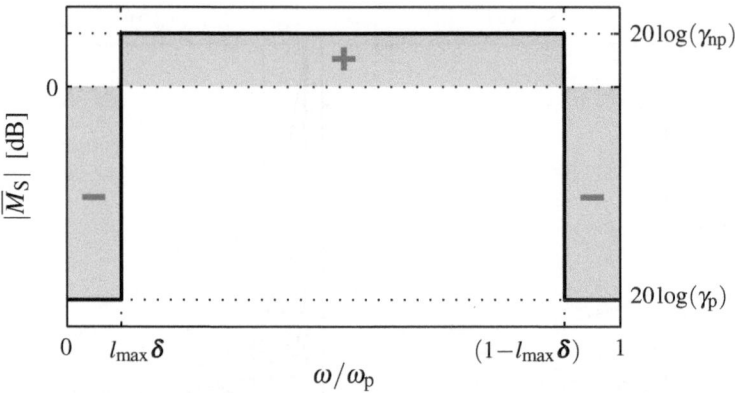

Fig. 5.8 Asymptotic behavior of $|\overline{M}_S(\omega)|$ for $\mu \to \infty$.

Theorem (5.13) dictates equality of the positive and negative gray-shaded areas, which mathematically yields:

$$\gamma_{np} = \exp\left(-\ln(\gamma_p)\frac{l_{max}\delta}{0.5 - l_{max}\delta}\right).$$

(5.15)

If T_p is known with infinite accuracy, $\delta = 0\%$, the trade-off between γ_p and γ_{np} vanishes for $\mu \to \infty$, as then, (5.15) yields $\gamma_{np} = 1$, independent of γ_p. The dotted lower bound for the trade-off curve between γ_p and γ_{np} is independent of the repetitive controller's parametrization, since for $\mu \to \infty$, the FIR parametrization (5.2) encompasses any stable transfer function $\chi(z)$ as a rational function of z^{-N}.

Figure 5.7 reveals that the performance gained by increasing the transient response time saturates: while initially, the trade-off curves shift significantly to the left and/or downwards as μ increases, for high μ values the trade-off curves lie closer and closer to each other. This saturation is observed for all $l_{max}\delta$ values.

To further analyze performance as a function of μ, Figure 5.9 shows the cross-sections through the trade-off surfaces of Figure 5.7 at $\gamma_{np} = 1.3$. Hence, this figure shows how the periodic performance improves as the transient response time increases. The dashed lines indicate the asymptotic γ_p values, predicted by (5.15). While Figure 5.9(a) reveals that for $\delta = 0\%$, the asymptotic value $\gamma_p = 0$ is reached with finite μ, for $\delta > 0\%$, the asymptotic γ_p value is only reached for $\mu \to \infty$, as observed in Figures 5.9(b), 5.9(c) and 5.9(d).

For $\mu < 20$, the curves of Figures 5.9(b), 5.9(c) and 5.9(d) feature alternating parts with steep and shallow steps. From an engineering point of view, μ values at the foot of a steep part are preferred: reducing μ would result in a substantial loss of performance, whereas little performance is gained by increasing μ. Figures 5.9(b), 5.9(c) and 5.9(d) confirm the saturation of the performance gained by increasing μ, and for $\mu > 20$, γ_p only decreases very slowly as μ increases. Hence, repetitive controllers with $\mu > 20$ are only relevant to applications with tight steady-state performance demands, while transient response time is of minor importance.

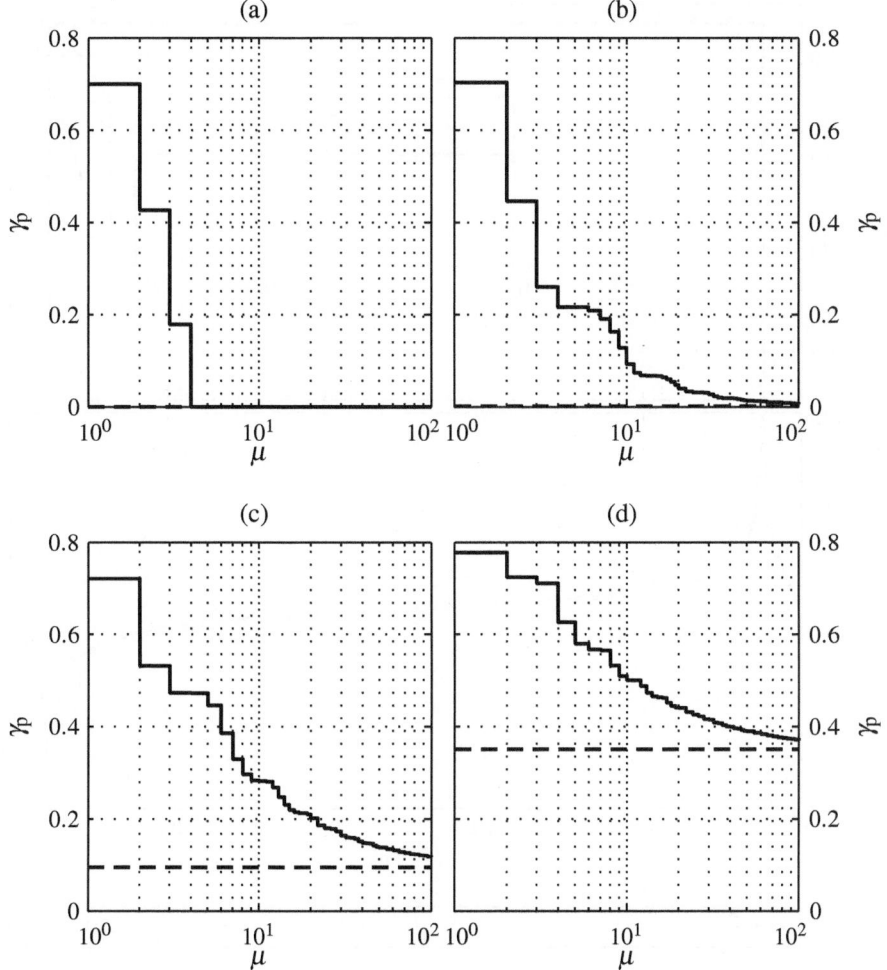

Fig. 5.9 Evolution of γ_p as a function of μ for $\gamma_{np} = 1.3$ and four uncertainty levels $l_{max}\delta$ on the fundamental frequency: (a) $l_{max}\delta = 0\%$; (b) $l_{max}\delta = 2\%$; (c) $l_{max}\delta = 5\%$; and (d) $l_{max}\delta = 10\%$. The dashed lines indicate the asymptotic γ_p values for $\mu \to \infty$.

Up to $\mu = 15$, problem (5.14) is most efficiently solved by combining the (generalized) KYP lemma and the solver of [94], while for higher μ, gridding becomes more efficient. For $\mu = 100$, the former approach requires about one CPU minute to solve (5.14), while SDPT3 [141, 149] solves the SOCP resulting from gridding within a few (< 5) CPU seconds (Intel® Core™2 Duo T9300, 2.5 GHz, 3.5 GB of RAM).

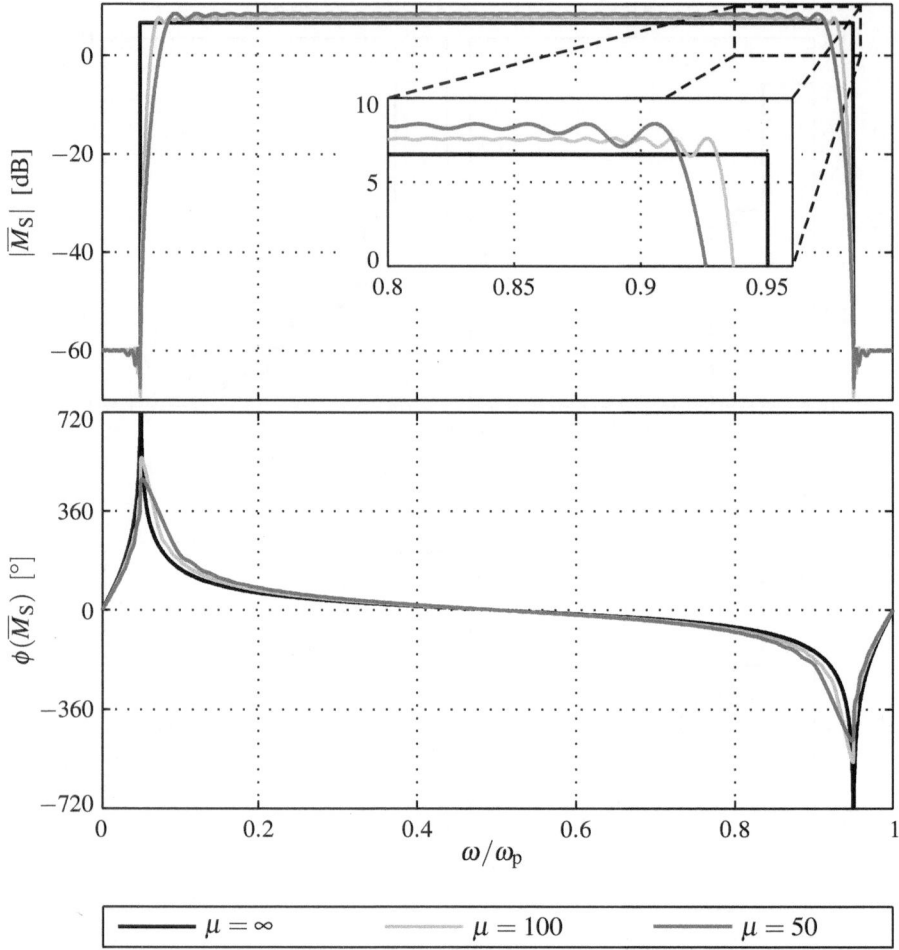

Fig. 5.10 FRF of $\overline{M}_S(z)$ for three repetitive controllers designed to yield $\gamma_p = 10^{-3}$ for $l_{max}\boldsymbol{\delta} = 5\%$.

Evolution of $\overline{M}_S(z)$ as a Function of μ

Figure 5.10 shows the FRF of $\overline{M}_S(z)$ for three repetitive controllers, designed to yield $\gamma_p = 10^{-3}$ for $l_{max}\boldsymbol{\delta} = 5\%$. The results are shown for $\mu = 50$, $\mu = 100$ and $\mu \to \infty$, where the latter result is constructed from Figure 5.8 by spectral factorization [107, 120]. As μ increases, $\overline{M}_S(z)$ indeed approaches the asymptotic behavior predicted in Figure 5.8. However, approximating the sharp edges in the asymptotic $|\overline{M}_S(\omega)|$ curve with a FIR parametrization gives rise to a Gibbs-like phenomenon [51].

Figure 5.11 shows the step response of $\overline{M}_S(z)$ for the repetitive controllers considered in Figure 5.10. Since $\overline{M}_S(z)$ only contains powers of z^{-N}, the step response

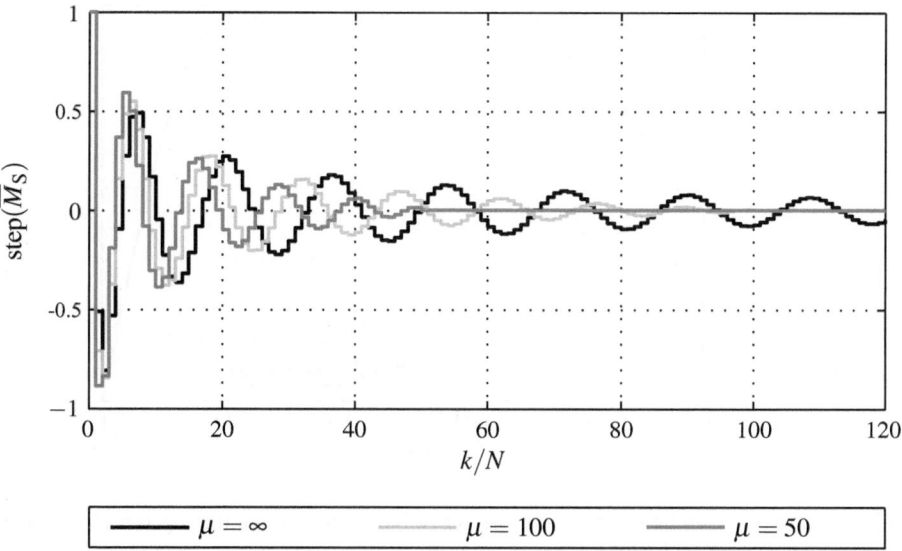

Fig. 5.11 Step response of $\overline{M}_S(z)$ for three repetitive controllers designed to yield $\gamma_p = 10^{-3}$ for $l_{max}\delta = 5\%$.

only changes at time instants k corresponding to a multiple of N. The step response of $\overline{M}_S(z)$ determines the transient response of all harmonic components, since, due to its FRF periodicity, $\overline{M}_S(z)$ treats all harmonics similar to a step input. For a μth-order repetitive controller, $\overline{M}_S(z)$ has a finite impulse response of length μN and consequently, steady state is reached after μ periods. Figure 5.11 reveals that for lower μ, the step response of $\overline{M}_S(z)$ is faster: the step response for $\mu \to \infty$ lags behind the response for $\mu = 100$, which is slower than the response for $\mu = 50$. For $\mu \to \infty$, the step response corresponds to a decaying harmonic signal, of which the period approaches $20N$, corresponding to the frequency δf_p. Figure 5.11 confirms that pushing the repetitive controller design to the asymptotic $\gamma_p - \gamma_{np}$ trade-off curve, comes at the price of a sluggish transient response: for $\mu \to \infty$, the transient response decays very slowly.

5.4.3 Comparison with the Literature

This section illustrates the capability of the proposed design approach to reproduce and outperform the current high-order repetitive controller designs of Chang *et al.* [20] and Steinbuch [135].

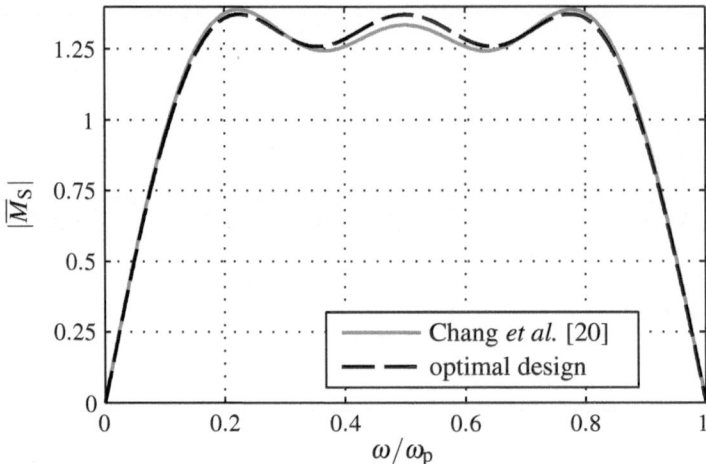

Fig. 5.12 Comparison between the optimal repetitive controller and the repetitive controller designed by Chang *et al.* [20] for $\mu = 3$.

Chang *et al.* [20]

To improve the nonperiodic performance of a repetitive controller that yields perfect nominal periodic performance, Chang *et al.* [20] design a high-order controller according to optimization problem (5.7). As already pointed out in [136], inequality constraints (5.7c) are superfluous. As a result, optimization problem (5.7) corresponds to the left-most point of the corresponding trade-off curve for $\delta = 0\%$, computed by solving (5.14) with a very small α value. Whereas Chang *et al.* [20] solve semi-infinite optimization problem (5.7) with a stochastic "evolution strategy", the proposed methodology provides a systematic and efficient approach to compute their solutions with guaranteed global optimality.

Figure 5.12 illustrates for $\mu = 3$ the difference between the optimal solution and the solution of [20]. Based on the guarantee of obtaining a global optimum, the optimal solution ($\gamma_{np} = 1.37$) is better, albeit only marginally, than the solution of [20] ($\gamma_{np} = 1.39$).

Steinbuch [135]

To improve robust periodic performance under the constraint of perfect nominal periodic performance, Steinbuch [135] enforces the higher-order derivatives of $|\overline{M}_S(\omega)|$ equal to zero at the multiples of ω_p (5.8). While this analytical design approach does not involve numerical optimization, the results can be approximated by solving optimization problem (5.14) with $\alpha = 0$ for a small uncertainty $l_{\max}\delta$. Figure 5.13 compares for $\mu = 3$ the result of [135] with the optimal solution of (5.14) with $\alpha = 0$ (corresponding to the left-most point of the trade-off

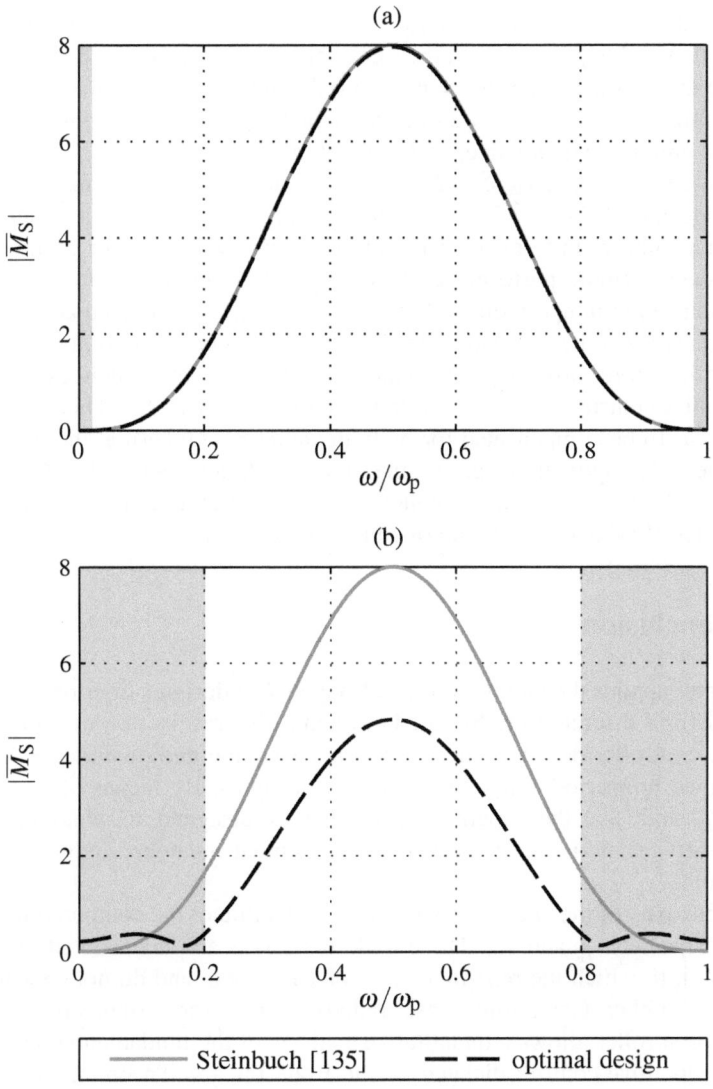

Fig. 5.13 Comparison of the optimal repetitive controller and the repetitive controller designed by Steinbuch [135] for $\mu = 3$ and two values of $l_{\max}\boldsymbol{\delta}$: (a) $l_{\max}\boldsymbol{\delta} = 2\%$; and (b) $l_{\max}\boldsymbol{\delta} = 20\%$. The shaded bands indicate the corresponding uncertainty intervals $\overline{\Omega}_{l_{\max}}$.

curve). Figures 5.13(a) and 5.13(b) respectively show the results for $l_{\max}\boldsymbol{\delta} = 2\%$ and $l_{\max}\boldsymbol{\delta} = 20\%$, where the corresponding uncertainty intervals $\overline{\Omega}_{l_{\max}}$ are shaded gray.

For small uncertainty, $l_{\max}\boldsymbol{\delta} = 2\%$, the results of the two design approaches almost coincide. Steinbuch [135] yields $\gamma_p = 2 \cdot 10^{-3}$ and $\gamma_{np} = 8$, while the optimal design yields $\gamma_p = 4.98 \cdot 10^{-4}$ and $\gamma_{np} = 7.96$. If an optimal repetitive controller is designed that yields $\gamma_p = 2 \cdot 10^{-3}$, the same value as [135], a better nonperiodic performance index is achieved: $\gamma_{np} = 6.97$, a reduction by 13% compared to $\gamma_{np} = 8$. However, since both solutions are situated to the left of knee x in Figure 5.6(b), none of them is preferable in practice, as discussed in Section 5.4.1.

For large uncertainty, $l_{max}\delta = 20\%$, the presented approach differs significantly from the design of Steinbuch [135], which, being derivative based, inherently assumes small uncertainty levels. The optimal repetitive controller design features both a better periodic performance index ($\gamma_p = 0.37$ compared to $\gamma_p = 1.62$ for [135]) and a better nonperiodic performance index ($\gamma_{np} = 4.83$ compared to $\gamma_{np} = 8$ for [135]). This can be understood from the Bode Integral Theorem (5.13) as constraints (5.8) enforce $|\overline{M}_S(\omega)| \approx 0$ around the multiples of ω_p, which results in a very large negative contribution of these frequencies in integral (5.13). This negative contribution has to be compensated for by high values of $|\overline{M}_S(\omega)|$ at the intermediate frequencies. The optimal design, on the other hand, does not push $|\overline{M}_S(\omega)|$ down to zero, yielding a substantially smaller negative contribution in (5.13), and hence, lower values of $|\overline{M}_S(\omega)|$ at the intermediate frequencies.

5.5 Conclusion

This chapter applies the methodology of Chapter 2 to design a high-order repetitive controller for a discrete-time SISO LTI system. This results in a novel high-order repetitive controller design, which allows a systematic and quantitative treatment of combined nonperiodic and period-uncertain inputs. By means of performance indices γ_p and γ_{np}, the repetitive controller is designed to yield an optimal trade-off between periodic performance improvement and nonperiodic performance degradation.

The convexity of the optimal design problem facilitates the computation of trade-off curves between γ_p and γ_{np}. These trade-off curves are independent of the feedback system to which the repetitive controller is added, and do not depend on the input period either. Computing these trade-off curves for various orders μ of the repetitive controller allows a quantitative analysis of the fundamental performance limits in repetitive control: dictated by the Bode Integral Theorem, any repetitive controller is bound to a trade-off between periodic performance improvement, nonperiodic performance degradation and transient response time.

The proposed design approach is able to reproduce and outperform the results of Chang et al. [20] and Steinbuch [135]. The high-order repetitive control design approach of [20] corresponds to the left-most point of the corresponding trade-off curve. The proposed repetitive controller design outperforms the approach of Steinbuch [135]: the same periodic performance is achieved with significantly less nonperiodic performance degradation, and this performance gain increases with period-time uncertainty.

Chapter 6
Application to Feedback Control

6.1 Introduction

6.1.1 State of the Art

This chapter deals with the general design of a feedback controller in the presence of periodic inputs. Repetitive controllers, discussed in the previous chapter, are the most popular type of feedback controllers for periodic inputs (see references in Section 5.1), where this popularity is to a large extent related to their simple structure, shown in Figure 5.2(b), and intuitive design. Moreover, a repetitive controller design only requires knowledge of the input period T_p, and its implementation is easily made adaptive for varying periods by adjusting $N = \mathrm{int}(T_p/T_s)$ [42, 139]. On the other hand, its particular structure impedes a repetitive controller to properly account for additional information on the periodic input spectrum as well as uncertainty on the input period. These disadvantages are caused by the period delays z^{-N} embedded in the repetitive controller structure. The FRF of z^{-N} is periodic with $\omega_p = 2\pi/(NT_s)$ and this FRF periodicity enforces an equal treatment of all harmonics l. Hence, a repetitive controller: (i) also accounts for harmonics l not present in the input ($l \notin \mathscr{L}$); (ii) cannot attribute different weights W_l to the harmonics $l \in \mathscr{L}$; and (iii) a robust design for uncertainty on T_p is usually overly robust at the lower harmonic frequencies, since the uncertainty intervals Ω_l (2.4) grow linearly with harmonic number l. These disadvantages restrict the ability of a repetitive controller to optimize closed-loop periodic performance, and the quantification of the resulting performance loss is valuable for a control engineer who has to decide whether the advantages of the repetitive controller structure pay off its disadvantages.

In feedback control, the Bode Integral Theorem [10, 21, 22, 49, 69, 138] dictates a fundamental trade-off between closed-loop periodic performance, nonperiodic performance and duration of transient response (see Section 5.4.2 and Section 6.5.2). Hence, the performance degradation caused by the repetitive controller structure should be analyzed in view of this trade-off and hereby invokes questions like: "For a given level of periodic and nonperiodic performance, how much is the transient

G. Pipeleers et al.: Optimal Linear Controller Design for Periodic Inputs, LNCIS 394, pp. 83–118.
springerlink.com   © Springer-Verlag Berlin Heidelberg 2009

response lengthened by adopting the repetitive controller structure?" or "For a given transient response time, how much performance can be gained by abandoning the repetitive controller structure?". To answer these questions, the fundamental performance trade-off in feedback control needs to be investigated in a quantitative manner, similar to the analysis of Section 5.4.2 for repetitive controllers.

The current literature results only allow for a partial analysis of this trade-off since the feedback controller design for periodic inputs is usually considered as an output regulation problem (see Appendix B for an introduction and [117] for an in-depth treatment). A regulator is a controller that achieves perfect asymptotic tracking/rejection of persistent inputs, and according to the Internal Model Principle [34, 45, 46, 47, 48] it includes a (partial) copy of the input signal generator. Hence, by reformulating the feedback controller design as a regulation problem, perfect periodic performance is inherent to the design. Moreover, uncertainty on the input period cannot be accounted for, since regulation theory cannot cope with uncertainty on the signal generator.

The (partial) copy of the signal generator is usually complemented with additional design freedom to optimize alternative performance specifications [1, 2, 3, 4, 19, 71, 118, 121, 133, 137]. Among these contributions, only Scherer *et al.* [121] allow investigating the remaining trade-off between nonperiodic performance and transient response time. The latter is controlled by constraining the closed-loop poles to the disc with radius $\beta < 1$ centered at the origin, which forces the transient response to decay with at least β^k. By means of the Lyapunov shaping paradigm, closed-loop nonperiodic performance can be optimized while satisfying this pole placement constraint [121]. Hillerström and Sternby [65] propose a similar feedback controller design as they place the closed-loop poles on the circle with radius $\beta < 1$ centered at the origin. Scalar β controls the trade-off between nonperiodic performance and the exponential decay rate of the transient response.

Recent research in output regulation deals with relaxing the perfect periodic performance resulting from the Internal Model Principle to good periodic performance [72, 86, 88, 93]. The result of Köroğlu and Scherer [88] is most relevant to this chapter as it extends the trade-off analysis by Scherer *et al.* [121] with this relaxation.

6.1.2   Contribution

This chapter applies the methodology of Chapter 2 to design a feedback controller for a discrete-time SISO LTI system. The design is similar to the repetitive controller design presented in Chapter 5 and shares its advantages:

Multi-objective Control: Contrary to the output regulation based design approaches, perfect periodic performance is not the starting-point of the presented feedback controller design. Instead, the controller is designed to yield an optimal trade-off between two performance indices, γ_p and γ_{np}, which quantify its effect on the closed-loop periodic and nonperiodic performance, respectively.

Period-time Uncertainty: Periodic performance index γ_p explicitly accounts for period-time uncertainty, whereas this uncertainty cannot be accounted for in output regulation.

Limits of Performance: The convex reformulation of the optimal design problem facilitates the computation of trade-off curves between the conflicting performance indices γ_p and γ_{np}. The length of the transient response, the third issue involved in the fundamental performance trade-off, is determined by the controller parametrization. This way, the presented feedback controller design allows a systematic and quantitative analysis of the fundamental limits of performance in feedback control.

In applying the general design methodology, the Youla parametrization is chosen such that repetitive controllers are encompassed as a special case. The resulting controller structure is therefore indicated as "generalized repetitive control" [111], and to emphasize the distinction between repetitive and generalized repetitive controllers, the former controllers are in this chapter referred to as "typical repetitive controllers". The property that typical repetitive controllers constitute a subclass of generalized repetitive controllers allows investigating the performance loss caused by the particular structure of the former type of controllers.

6.1.3   Outline

Section 6.2 details the control setup used in this chapter and presents the structure of a generalized repetitive controller. This section also reviews the feedback controller designs of Hillerström and Sternby [65], Scherer *et al.* [121] and Köroğlu and Scherer [88]. Section 6.3 clarifies how the generalized repetitive controller structure encompasses typical repetitive controllers as a special case and reveals a close relationship between generalized repetitive control and estimated disturbance feedback control, presented in Chapter 4. Section 6.4 applies the developed methodology to the generalized repetitive controller design and elaborates on the corresponding general control configuration, Youla parametrization and optimal design, while Section 6.5 illustrates its potential by numerical results. Section 6.6 summarizes the conclusions of this chapter.

6.2   Background

Section 6.2.1 details the control configuration used in this chapter and presents the structure of a generalized repetitive controller. Subsequently, Section 6.2.2 briefly reviews the current feedback controller designs [65, 88, 121].

6.2.1   Control Configuration

To enhance the similarity with the previous chapter, the feedback controller, $K_{\text{FB}}(z)$, is considered as an add-on device, that is: it is added to the loop of an existing feedback system, referred to as the "original feedback system", to improve its closed-loop periodic performance. The original feedback system comprises plant $G(z)$ and original feedback controller $K_o(z)$, which has been designed *a priori* and is hence considered fixed. Figure 6.1(a) shows the corresponding control setup, where signals $r(k)$, $d(k)$ and $\eta(k)$ respectively correspond to the reference input, output disturbance and plant output. In closed loop, the tracking error $e(k) = r(k) - \eta(k)$ is given by

$$e(k) = S(q)\big(r(k) - d(k)\big) ,$$

where $S(q)$ corresponds to the closed-loop sensitivity. The combined input $r(k) - d(k)$ features both a periodic and a nonperiodic contribution:

$$r(k) - d(k) = w_{\text{p}}(k) + w_{\text{np}}(k) , \tag{6.1}$$

where $w_{\text{p}}(k)$ is specified according to Section 2.2.2. Contrary to the previous chapter, the period T_{p} is not required to contain an integer number of sample periods.

Original Feedback System

$K_o(z)$ must yield an internally stable closed-loop system and is hence indispensable for unstable plants $G(z)$. The sensitivity and complementary sensitivity function of the original feedback system are respectively denoted by $S_o(z)$ and $T_o(z)$:

$$S_o(z) = \frac{1}{1 + K_o(z)G(z)} , \qquad T_o(z) = \frac{K_o(z)G(z)}{1 + K_o(z)G(z)} .$$

For ease of explanation, $K_o(z)$ is assumed to be stable and designed properly, which implies [131]: (i) a stable closed-loop system; (ii) high-gain feedback at low frequencies; (iii) sufficient roll-off of $|T_o(\omega)|$ at high frequencies; and (iv) a large modulus margin $\|S_o(z)\|_\infty^{-1}$. These assumptions are also made in Chapter 5, and Figure 5.3 illustrates the FRFs of $S_o(z)$ and $T_o(z)$ corresponding to a proper design of $K_o(z)$.

   Property (ii) is referred to as good nonperiodic performance, since it yields small $|S_o(\omega)|$ at low frequencies, resulting in a small tracking error $e(k)$ for any low-frequency input $w_{\text{np}}(k)$. By defining the bandwidth ω_{BW} of the original feedback system as the frequency where $|S_o(\omega)|$ first crosses $-3\,\text{dB}$ from below, see Figure 5.3(a), $K_o(z)$ is said to yield good nonperiodic performance up to ω_{BW}.

   Combination of properties (i), (iii) and (iv) ensures robust stability of the original feedback system. In the presence of multiplicative unstructured plant uncertainty (2.2), robust stability of the original feedback system requires $\|T_o(z)W_G(z)\|_\infty < 1$ [131], where optimizing closed-loop performance pushes $|T_o(\omega)|$ at high

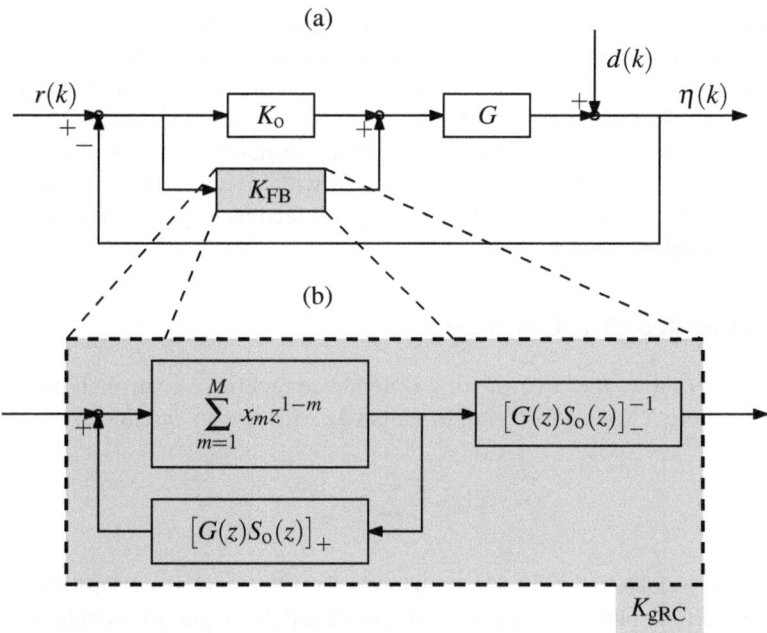

Fig. 6.1 (a) Add-on feedback control configuration, where controller K_{FB} is added to the "original feedback system" comprising plant G and controller K_o; and (b) structure of a generalized repetitive controller K_{gRC}. Signals $r(k)$, $d(k)$ and $\eta(k)$ respectively denote the reference trajectory, output disturbance and plant output, while S_o indicates the sensitivity function of the original feedback system.

frequencies to its upper bound $|1/W_G(\omega)|$, as illustrated in Figure 5.3(b). The modulus margin $\|S_o(z)\|_\infty^{-1}$ corresponds to the minimal distance between the Nyquist plot of the loop transfer function and the point -1, and is therefore also considered as a robust stability measure.

Overall Feedback System

When $K_{FB}(z)$ is added to the loop, the closed-loop sensitivity changes from $S_o(z)$ to $S(z)$:

$$S(z) = \frac{1}{1 + \left[K_o(z) + K_{FB}(z)\right]G(z)},$$

$$= S_o(z) \underbrace{\frac{1}{1 + K_{FB}(z)G(z)S_o(z)}}_{M_S(z)}.$$

Transfer function $M_S(z)$ is called the modifying sensitivity function and represents the effect of $K_{FB}(z)$ on the closed-loop sensitivity. $K_{FB}(z)$ must not compromise robust stability of the original feedback system, while improving closed-loop periodic performance. These design specifications for $K_{FB}(z)$ are nonconflicting provided that all harmonics $l \in \mathscr{L}$ lie well below ω_{BW}: at harmonics near or above ω_{BW} better performance simply cannot be achieved without compromising the modulus margin or high-frequency roll-off of the original feedback system. This assumption is made throughout this chapter.

Generalized Repetitive Controller

Figure 6.1(b) shows the structure of a generalized repetitive controller, indicated by $K_{gRC}(z)$, where $X(z)$ is a FIR filter of length M with design variables x_m:

$$X(z) = \sum_{m=1}^{M} x_m z^{1-m} . \tag{6.2}$$

The invertible part of $G(z)S_o(z)$ is denoted by $\left[G(z)S_o(z)\right]_-$, where the remaining noninvertible part $\left[G(z)S_o(z)\right]_+$ comprises a delay equal to the relative degree of $G(z)S_o(z)$ and its nonminimum-phase zeros. The nonminimum-phase zeros of $G(z)S_o(z)$ only stem from $G(z)$, since the nonminimum-phase zeros of $S_o(z)$ correspond to the unstable system poles and consequently cancel in the multiplication with $G(z)$. If $G(z)$ and/or $K_o(z)$ is strictly causal, $S_o(z)$ has zero relative degree, whereby

$$\left[G(z)S_o(z)\right]_- = G_-(z)S_o(z) , \tag{6.3a}$$

$$\left[G(z)S_o(z)\right]_+ = G_+(z) . \tag{6.3b}$$

The controller structure of Figure 6.1(b) gives rise to the following expressions for $K_{gRC}(z)$ and $M_S(z)$:

$$K_{gRC}(z) = \left[G(z)S_o(z)\right]_-^{-1} \frac{X(z)}{1 - \left[G(z)S_o(z)\right]_+ X(z)} , \tag{6.4a}$$

$$M_S(z) = 1 - \left[G(z)S_o(z)\right]_+ X(z) . \tag{6.4b}$$

6.2.2 Current Design Approaches

This section briefly reviews the current feedback controller designs of Hillerström and Sternby [65], Scherer *et al.* [121] and Köroğlu and Scherer [88]. These design strategies adopt a controller structure that differs from the generalized repetitive controller structure of Figure 6.1(b).

Hillerström and Sternby [65]

Hillerström and Sternby [65] present a discrete-time SISO LTI feedback controller
design for periodic inputs, and although they don't explicitly adopt the add-on control configuration of Figure 6.1(a), this setup is beneficial if both nonperiodic and
periodic inputs enter the control loop.

A feedback controller designed according to [65] comprises three parts: (i)
through pole-zero cancelation, the undesired plant dynamics are eliminated from
the closed-loop system; (ii) dictated by the Internal Model Principle, the controller
includes signal generator $\Lambda(z)$, Equation 2.5, to achieve perfect asymptotic track-
ing/rejection of all harmonics $l \in \mathscr{L}$; and (iii) an additional controller part assigns
the closed-loop poles to the poles of $\Lambda(z)$ (which lie on the unit circle) multiplied
by a positive scalar $\beta < 1$. As illustrated in Section 6.5.3, β governs the trade-off
between nonperiodic performance and transient response time. Related to the Internal Model Principle, perfect periodic performance is intrinsic to the design, and
uncertainty on the input period cannot be accounted for.

Translated to the add-on control configuration of Figure 6.1(a), the design approach of [65] yields

$$K_{\text{FB}}(z) = \left[G(z)S_{\text{o}}(z) \right]_-^{-1} \Lambda(z) K_{\text{pp}}(z) \, ,$$

where $K_{\text{pp}}(z)$ is the controller part that places the closed-loop poles at their desired
locations.

Scherer *et al.* [121]; Köroğlu and Scherer [88]

A feedback controller designed according to Scherer *et al.* [121] comprises two
parts: (i) a copy of signal generator $\Lambda(z)$, Equation 2.5, which guarantees perfect periodic performance; and (ii) a complementary controller that stabilizes the
closed-loop system and provides the design freedom to optimize alternative performance specifications. The design of this complementary controller is facilitated by
the derivation of an auxiliary plant (see Appendix B), and relies on the Lyapunov
shaping paradigm [121]. This paradigm provides an approximate, convex approach
to design a controller according to multiple specifications that can be recast into
LMIs (see e.g. [14, 121] for an overview).

The additional design specifications of interest are nonperiodic performance and
transient response speed. In this monograph, nonperiodic performance is quantified
by γ_{np} (2.10), which equals the $\mathscr{H}_\infty$ norm of $H_{\text{np}}(z)$ and complies with the Lyapunov
shaping paradigm. Although the closed-loop transient response can be designed rig-
orously via $\mathscr{H}_2$ techniques, the resulting controller design generally depends on the
periodic input and the system's initial conditions [118]. Pole placement is a more
elegant way to control the transient response speed, since constraining the closed-
loop poles to the disc with radius $\beta < 1$ centered at the origin forces the transient
response to decay with at least β^k. By combining this pole placement constraint

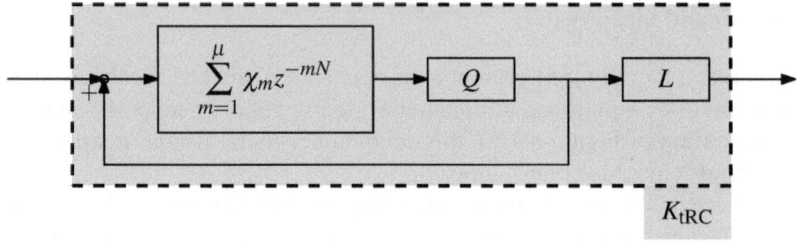

Fig. 6.2 Structure of a μth-order typical repetitive controller K_{tRC}.

with γ_{np}, Scherer *et al.* [121] allow investigating the trade-off between transient response time and nonperiodic performance under the constraint of perfect periodic performance.

Köroğlu and Scherer [88] extend the above described analysis with the possibility to relax the perfect periodic performance constraint. Their modification only involves the controller part related to the Internal Model Principle, while the design of the complementary controller part is preserved from [121]. Instead of including a copy of $\Lambda(z)$, this controller part is modified (see Appendix B) to guarantee that in steady state:

$$|v_p(k)| \leq \kappa \, \mathrm{rms}(w_p(k)) \,, \tag{6.5}$$

for any periodic input $w_p(k)$, where $v_p(k)$ indicates the regulated output related to closed-loop periodic performance. The design approach of Köroğlu and Scherer [88] encompasses the result of Scherer *et al.* [121] by setting $\kappa = 0$, while uncertainty on the input period cannot be accounted for.

6.3   Relation with Alternative Controller Structures

This section illustrates how typical repetitive controllers correspond to a subclass of generalized repetitive controllers (Section 6.3.1), and reveals a close relationship between generalized repetitive control and estimated disturbance feedback control, presented in Chapter 4 (Section 6.3.2).

6.3.1   Relation with Typical Repetitive Control

Figure 6.2 resumes the structure of a μth-order typical repetitive controller of Figure 5.2(b). A typical repetitive controller relies on the internal model shown in Figure 5.1(b), where

$$N = \mathrm{int}\left(T_p/T_s\right) \,,$$

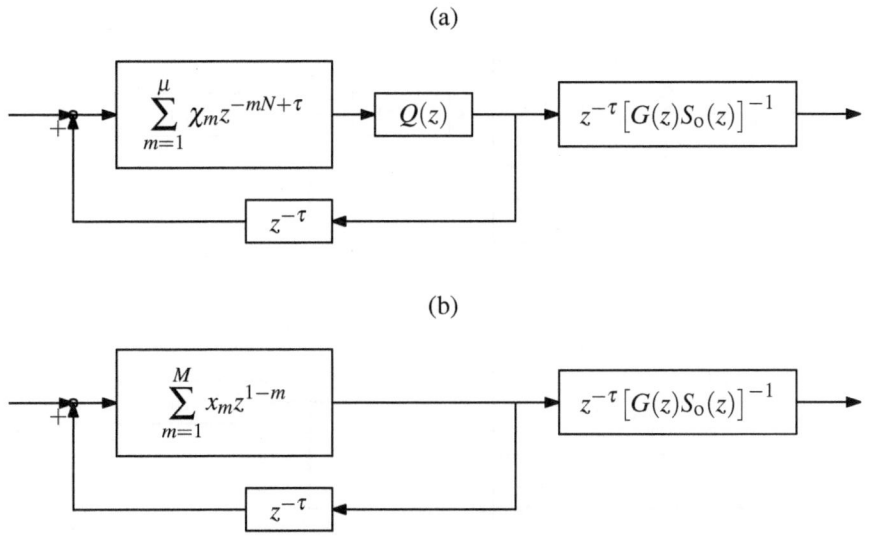

Fig. 6.3 Structure of a typical repetitive controller K_{tRC} (a) and a generalized repetitive controller K_{gRC} (b) in the case of a minimum-phase system G, where $\tau \geq 0$ denotes the relative degree of $G(z)S_o(z)$.

and the rounding error is accounted for as additional uncertainty on $T_p \equiv NT_s$. In high-order typical repetitive controllers, the delay z^{-N} is extended to a polynomial in z^{-N}:

$$\chi(z) = \sum_{m=1}^{\mu} \chi_m z^{-mN} \,,$$

which gives the designer more freedom to improve closed-loop performance through a proper design of the parameters χ_m (see Chapter 5). $L(z)$ guarantees nominal stability of the closed-loop system, whereas low-pass filter $Q(z)$ improves its robust stability (see Section 5.2.2).

To reveal the similarity between typical and generalized repetitive controllers, Figure 6.3 compares both structures for a minimum-phase system $G(z)$, where $\tau \geq 0$ denotes the relative degree of $G(z)S_o(z)$. In this case, $L(z) = \left[G(z)S_o(z)\right]^{-1}$ is the common design procedure in typical repetitive control, where noncausality of $L(z)$ due to $\tau > 0$, is accounted for as illustrated in Figure 6.3(a). For a minimum-phase system $G(z)$: $\left[G(z)S_o(z)\right]^{-1}_+ = z^{-\tau}$ and $\left[G(z)S_o(z)\right]^{-1}_- = z^{-\tau}\left[G(z)S_o(z)\right]^{-1}$, whereby the generalized repetitive controller structure amounts to Figure 6.3(b). Typical repetitive controllers constitute a subclass of generalized repetitive controllers since a FIR filter design of $Q(z)$, as is common practice in typical repetitive control [28, 61, 136], turns $\chi(z)Q(z)$ into a FIR filter with a particular structure.

For a nonminimum-phase system $G(z)$, in typical repetitive control $L(z)$ is generally not designed as $\left[G(z)S_o(z)\right]^{-1}_-$, but set equal to an approximate, stable inverse of

(a)

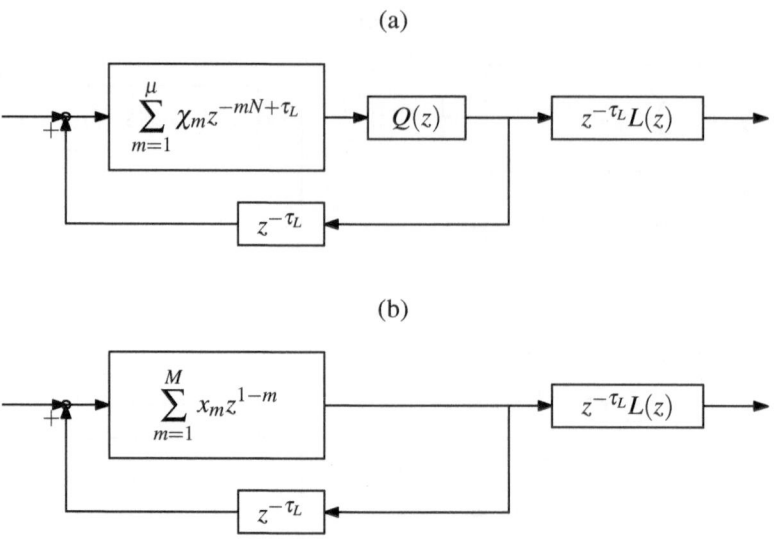

(b)

Fig. 6.4 Structure of a typical repetitive controller K_{tRC} (a) and alternative structure of a generalized repetitive controller K_{gRC} (b) in the case of a nonminimum-phase system G, where $L(z)$ corresponds to an approximate stable inverse of $G(z)S_o(z)$, while τ_L equals the smallest integer that renders $z^{-\tau_L}L(z)$ causal.

$[G(z)S_o(z)]$. The "zero phase error tracking" inversion [142] is often used [61, 146], while alternative approximations are indicated in Section 3.1. These approximations are generally noncausal, whereby the smallest integer τ_L that renders $z^{-\tau_L}L(z)$ causal is larger than τ, and this noncausality is accounted for as indicated in Figure 6.4(a). With this design of $L(z)$, the typical repetitive controller no longer corresponds to a particular generalized repetitive controller. However, if this property is crucial, for a nonminimum-phase system $G(z)$ the generalized repetitive controller structure of Figure 6.1(b) can be modified to the one of Figure 6.4(b). This chapter continues with the former structure since it does not require approximate inversion techniques, while the elaboration for Figure 6.4(b) is very similar.

6.3.2 Relation with Estimated Disturbance Feedback Control

Figure 6.5 shows an equivalent representation of the estimated disturbance feedback control system of Figure 4.1(b). It comprises add-on feedback controller $K_{FB}(z)$:

$$K_{FB}(z) = \frac{S_o(z)^{-1}K_{dFB}(z)}{1 - G(z)K_{dFB}(z)},\qquad(6.6)$$

and feedforward controller $K_{FF}(z)$:

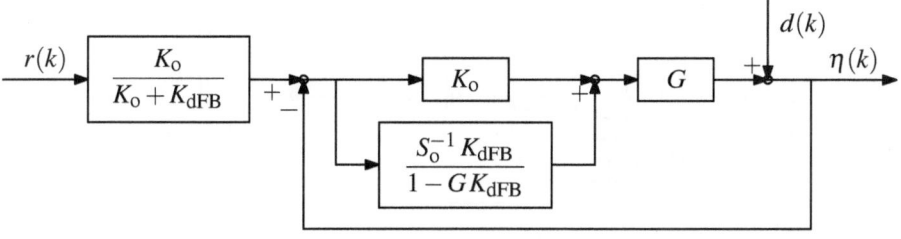

Fig. 6.5 Equivalent representation of the estimated disturbance feedback control configuration shown in Figure 4.1(b).

$$K_{FF} = \frac{K_o(z)}{K_o(z) + K_{dFB}(z)} \, ,$$

which cancels the effect of the disturbance feedback controller on $r(k)$. Comparison between (6.6) and (6.4a) reveals that the generalized repetitive controller yields the same feedback controller as an estimated disturbance feedback controller with

$$K_{dFB}(z) = S_o(z) \big[G(z) S_o(z) \big]_-^{-1} X(z) \,.$$

On account of relations (6.3), this expression generally reduces to $K_{dFB}(z) = G_-(z)^{-1} X(z)$, which is also suggested in Equation 4.5. Hence, the major difference between generalized repetitive control and estimated disturbance feedback control is the absence of $K_{FF}(z)$. From this observation, the design guideline is extracted that when adding a feedback controller to the loop to improve closed-loop periodic performance, the effect of this controller on measurable, nonperiodic inputs should be canceled by feedforward control.

6.4 Application of the Design Methodology

This section applies the general methodology of Chapter 2 to design an add-on feedback controller $K_{FB}(z)$. Section 6.4.1 presents the corresponding general control configuration, while Section 6.4.2 shows that the generalized repetitive controller structure of Figure 6.1(b) corresponds to a particular Youla parametrization for this control problem. Section 6.4.3 presents the resulting optimal design problem.

While $K_{gRC}(z)$ constitutes a particular Youla parametrization for add-on feedback controllers $K_{FB}(z)$, it should be noted that the combination of $K_o(z)$ and $K_{gRC}(z)$ corresponds to a particular Youla parametrization for the set of internally stabilizing feedback controllers. Hence, no conservatism is introduced by adopting the add-on control configuration of Figure 6.1(a): for $M \to \infty$, Figure 6.1 can generate any internally stable closed-loop system.

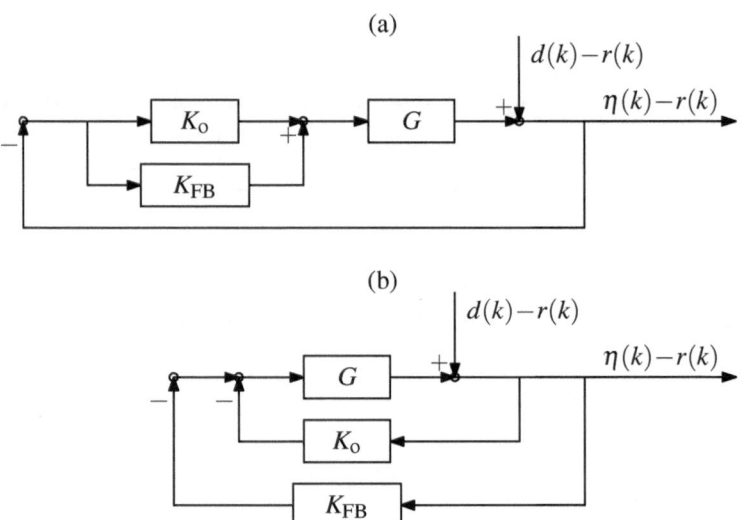

Fig. 6.6 Equivalent representations of the add-on feedback control configuration of Figure 6.1(a).

6.4.1   General Control Configuration

To introduce the derivation of the general control configuration related to Figure 6.1(a), this control configuration is first transformed into the equivalent schemes shown in Figure 6.6. Figure 6.6(a) is obtained from Figure 6.1(a) by shifting input $r(k)$ to the right-hand side, while Figure 6.6(a) is further simplified to Figure 6.6(b).

The design of $K_{FB}(z)$ faces three specifications: (i) it must not compromise the original feedback system's robust stability, while (ii) improving the closed-loop periodic performance at (iii) an acceptable nonperiodic performance degradation. These specifications are labeled i_{rs}, i_p and i_{np}, respectively, while notation $(\cdot)_{i_{rs}}$ is shortened to $(\cdot)_{rs}$, similar to notation $(\cdot)_p$ and $(\cdot)_{np}$. Figure 6.7 shows the corresponding general control configuration, where the three design specifications are related to the closed-loop subsystems from exogenous inputs $w_{rs}(k)$, $w_p(k)$ and $w_{np}(k)$ to regulated outputs $v_{rs}(k)$, $v_p(k)$ and $v_{np}(k)$, respectively.

To preserve robust stability, the overall feedback system should revert to the original feedback system for $\omega \geq \omega_{BW}$. This requirement implies that for $\omega \geq \omega_{BW}$,

$$|T(\omega) - T_o(\omega)| = |S_o(\omega) - S(\omega)| = |S_o(\omega)[1 - M_S(\omega)]|$$

must be small. By choosing $w_{rs}(k)$ and $v_{rs}(k)$ as indicated in Figure 6.7(a),

$$H_{rs}(z) = M_S(z) - 1 = \frac{-K_{FB}(z)S_o(z)G(z)}{1 + K_{FB}(z)S_o(z)G(z)},$$

and hence, robust stability is preserved if $|H_{rs}(\omega)|$ is small for all ω in $[\omega_{BW}, \pi f_s]$.

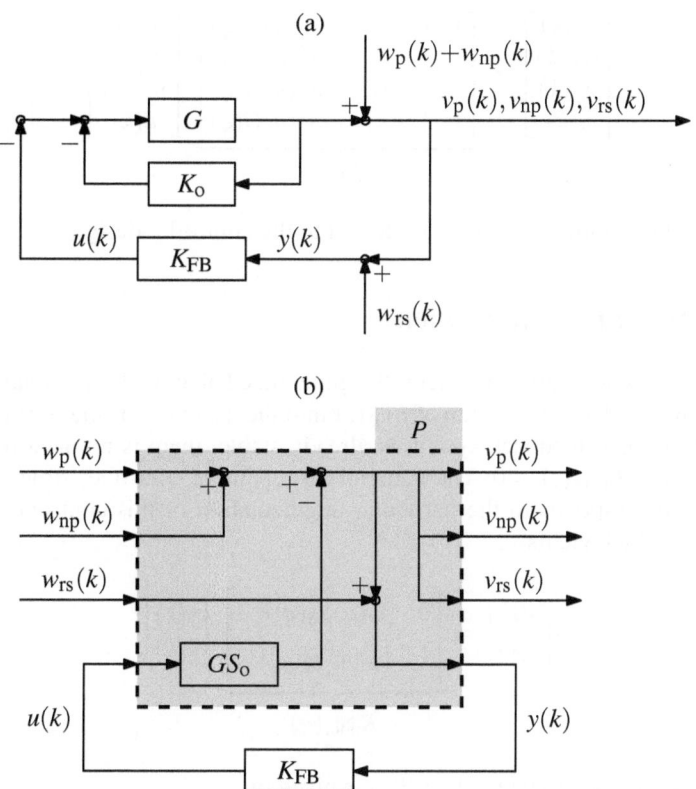

Fig. 6.7 General control configuration (b) for the design of add-on feedback controller K_{FB} (a); the design specifications involve the closed-loop subsystems from exogenous inputs $w_p(k)$, $w_{np}(k)$ and $w_{rs}(k)$ to regulated outputs $v_p(k)$, $v_{np}(k)$ and $v_{rs}(k)$, respectively.

The second and third design specification for $K_{FB}(z)$ relate to performance, and to focus on the effect of $K_{FB}(z)$ on the closed-loop performance, $H_p(z) = H_{np}(z) = M_S(z)$ is preferred over $S(z)$. These transfer functions are obtained by choosing exogenous inputs $w_p(k)$, $w_{np}(k)$ and regulated outputs $v_p(k)$, $v_{np}(k)$ as indicated in Figure 6.7(a). While substituting (6.1) in Figure 6.6(b) would yield $S(z)$ for $H_p(z)$ and $H_{np}(z)$, the modifying sensitivity is obtained by shifting the inputs beyond the branch to $K_o(z)$.

Figure 6.7(b) shows the internal structure of the resulting generalized plant $P(z)$, which yields

$$
\underbrace{\begin{bmatrix} v_p(k) \\ v_{np}(k) \\ v_{rs}(k) \\ y(k) \end{bmatrix} = \begin{bmatrix} 1 & \star & \star & -S_o(q)G(q) \\ \star & 1 & \star & -S_o(q)G(q) \\ \star & \star & 0 & -S_o(q)G(q) \\ 1 & 1 & -1 & -S_o(q)G(q) \end{bmatrix}}_{P(q)} \begin{bmatrix} w_p(k) \\ w_{np}(k) \\ w_{rs}(k) \\ u(k) \end{bmatrix} ,
$$

where symbol $\star$ indicates entries irrelevant to the controller design.

6.4.2 Youla Parametrization

The Youla parametrization augments the generalized plant with a nominal controller and hereby translates the design of $K_{FB}(z)$ into the design of Youla parameter $X(z)$. Since the original feedback system is already stable, there is no need for a nominal controller: $K_{nom}(z) = 0$. The generalized repetitive controller structure of Figure 6.1(b) corresponds to the particular augmentation of this controller shown in Figure 6.8, which yields

$$
\begin{bmatrix} u(k) \\ \tilde{y}(k) \end{bmatrix} = \underbrace{\begin{bmatrix} 0 & \left[G(q)S_o(q) \right]_-^{-1} \\ 1 & \left[G(q)S_o(q) \right]_+ \end{bmatrix}}_{\widetilde{K}_{nom}(q)} \begin{bmatrix} y(k) \\ \tilde{u}(k) \end{bmatrix} ,
$$

and the resulting augmented plant $\widetilde{P}(z)$ is given by

$$
\underbrace{\begin{bmatrix} v_p(k) \\ v_{np}(k) \\ v_{rs}(k) \\ \tilde{y}(k) \end{bmatrix} = \begin{bmatrix} 1 & \star & \star & -\left[G(q)S_o(q) \right]_+ \\ \star & 1 & \star & -\left[G(q)S_o(q) \right]_+ \\ \star & \star & 0 & -\left[G(q)S_o(q) \right]_+ \\ 1 & 1 & -1 & 0 \end{bmatrix}}_{\widetilde{P}(q)} \begin{bmatrix} w_p(k) \\ w_{np}(k) \\ w_{rs}(k) \\ \tilde{u}(k) \end{bmatrix} .
$$

As shown in Figure 6.8, Youla parameter $X(z)$ acts as a feedback controller for the augmented plant, and yields the following closed-loop transfer functions:

$$
\begin{aligned}
H_p(z) &= 1 - \left[G(z)S_o(z) \right]_+ X(z) , \\
H_{np}(z) &= 1 - \left[G(z)S_o(z) \right]_+ X(z) , \\
H_{rs}(z) &= - \left[G(z)S_o(z) \right]_+ X(z) .
\end{aligned}
$$

In the generalized repetitive controller, parametrization (2.17) is used for $X(z)$, see Equation 6.2, where the design parameters x_m are grouped in the vector $x \in \mathbf{R}_M$, Equation 2.16. By parameterizing $X(z)$ as a FIR filter of length M, transfer functions

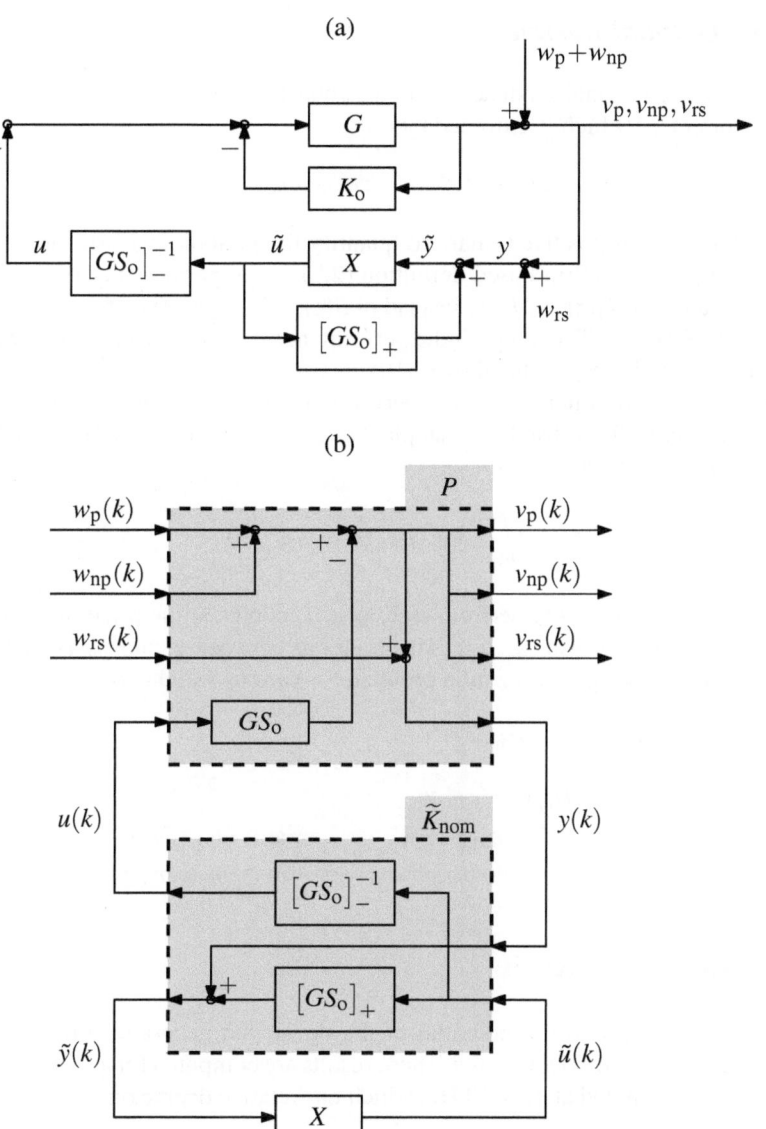

Fig. 6.8 Youla parametrization: augmentation of nominal controller $K_{nom} = 0$ corresponding to the generalized repetitive controller structure: (a) in the classical feedback control configuration; and (b) in the general control configuration. To save space, argument (k) of the sampled time signals is omitted in (a).

$H_p(z)$, $H_{np}(z)$ and $H_{rs}(z)$ have a finite impulse response as well, and their FIR lengths depend affinely on M. Hence, M determines the duration of the closed-loop transient response.

6.4.3   Optimal Design

Preserving the original feedback system's robust stability requires $|H_{rs}(\omega)|$ to be small from ω_{BW}, which is enforced by

$$|H_{rs}(\omega)| \leq \varepsilon, \quad \forall \omega \in [\omega_{BW}, \pi f_s],$$

where ε is a small positive scalar. To quantify the periodic performance improvement by $K_{FB}(z)$, ∞-norm based definition (2.7) of the periodic performance index γ_p is adopted. As $H_p(z) = M_S(z)$ instead of $S(z)$, it is appropriate to change weights W_l into $W_l|S_o(l\omega_p)|$. The nonperiodic performance degradation caused by $K_{FB}(z)$ is quantified by index γ_{np}, defined by (2.10).

For a given FIR filter length M, performance indices γ_p and γ_{np} are generally conflicting, since for a strictly causal plant $G(z)$ the Bode Integral Theorem [10, 21, 22, 49, 69, 138] dictates

$$\int_0^{\pi f_s} \log\left(|M_S(\omega)|\right) d\omega = 0. \tag{6.7}$$

Hence, improved periodic performance, $\gamma_p < 1$, comes at the price of nonperiodic performance degradation, $\gamma_{np} > 1$. This trade-off between γ_p and γ_{np} is analyzed by solving the following optimization problem for various weights $\alpha \geq 0$:

$$\underset{x, \gamma_p, \gamma_{np}}{\text{minimize}} \quad \gamma_p + \alpha \gamma_{np} \tag{6.8a}$$

$$\text{subject to} \quad \|H_{np}(z)\|_\infty \leq \gamma_{np} \tag{6.8b}$$

$$W_l|H_p(\omega)| \leq \gamma_p, \qquad \forall \omega \in \Omega_l, \quad \forall l \in \mathscr{L} \tag{6.8c}$$

$$|H_{rs}(\omega)| \leq \varepsilon, \qquad \forall \omega \in [\omega_{BW}, \pi f_s]. \tag{6.8d}$$

6.5   Numerical Results

This section illustrates the potential of the developed generalized repetitive controller design by numerical results. These results are computed for a minimum-phase system $G(z)$, sampled at $f_s = 1$ kHz, which has relative degree one:

$$\left[G(z)S_o(z)\right]_+ = G_+(z) = z^{-1}. \tag{6.9}$$

As revealed by (6.4b), this is the only plant information affecting the generalized repetitive controller design. The original feedback controller yields $\omega_{BW} = 2\pi 180$ rad/s, and its robust stability is preserved by enforcing constraint (6.8d) with $\varepsilon = 10^{-3}$. Periodic input $w_p(k)$ has nominal period $T_p = 0.05$ s, which corresponds to $f_p = 20$ Hz, and the generalized repetitive controller design accounts for harmonics $l \in \mathscr{L} = \{0, 1, 3, 5, 7\}$ with equal weight: $W_l = 1$, $\forall l \in \mathscr{L}$. These weights are extracted from Figure 3.4(b), where $S_o(z)$ is assumed to satisfy $W_l \equiv W_l|S_o(l\omega_p)| = 1$. Various uncertainty levels δ on f_p will be considered.

As discussed in Section 6.3.1, typical repetitive controllers constitute a subclass of generalized repetitive controllers. Section 6.5.1 compares both controller designs for the considered simulation example and investigates control problems for which typical repetitive controllers are optimal.

While for given M, the Bode Integral Theorem (6.7) implies a conflict between performance indices γ_p and γ_{np}, this theorem dictates a more fundamental trade-off in feedback control between periodic performance improvement, nonperiodic performance degradation and transient response time. The generalized repetitive controller design translates this trade-off into a trade-off surface between γ_p, γ_{np} and M. Section 6.5.2 analyzes this trade-off surface for the considered simulation example and investigates the solution of (6.8) for $M \rightarrow \infty$.

Subsequently, Section 6.5.3 compares the generalized repetitive controller design with the current feedback controller designs of Hillerström and Sternby [65] and Köroğlu and Scherer [88].

6.5.1 Generalized Versus Typical Repetitive Control

This section investigates the performance loss caused by the particular structure of a typical repetitive controller (Figure 6.2). To this end, typical repetitive controllers are compared to generalized repetitive controllers that yield the same impulse response length for $M_S(z)$, and hence, an equal transient response time. The following section complements the results presented here by comparing typical and generalized repetitive controllers that yield similar performance but different transient response times.

For the considered simulation example, a generalized repetitive controller with given M, yields a finite impulse response for $M_S(z)$ of $M + 1$ samples, as is clarified by the substitution of (6.9) in Equation 6.4b. A μth-order typical repetitive controller corresponds to a particular generalized repetitive controller with

$$M = \mu N + n_Q/2 , \tag{6.10}$$

where $N = \text{int}(T_p/T_s) = 50$, while n_Q denotes the order of FIR filter $Q(z)$. To obtain a fair comparison between generalized and typical repetitive controllers, $Q(z)$ is designed as the lowest-order zero-phase FIR filter for which all harmonics $l \in \mathscr{L}$ lie in its pass-band ($l_{max} = \max_{l \in \mathscr{L}}\{l\}$):

$$1 - 10^{-3} \leq |Q(\omega)| \leq 1 + 10^{-3} , \quad \forall \omega \in [0, l_{max}\omega_p] ,$$

while $|Q(\omega)| \leq \varepsilon$ for all ω in $[\omega_{BW}, \pi f_s]$, similar to (6.8d). This way, $n_Q = 88$. In this section, optimal typical and generalized repetitive controllers that correspond to a given M value, and hence, yield FIR length $M + 1$ for $M_S(z)$, are indicated by $K_{tRC,M}(z)$ and $K_{gRC,M}(z)$, respectively.

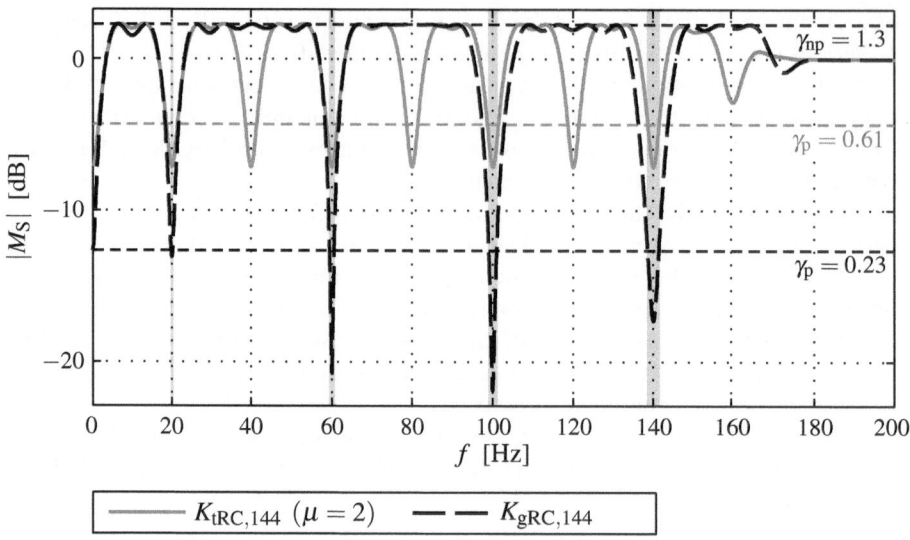

Fig. 6.9 Comparison of $|M_S(\omega)|$ for the $K_{tRC,149}(z)$ and $K_{gRC,149}(z)$ that yield $\gamma_{np} = 1.3$. The $\delta = 1\%$ uncertainty on f_p results in the gray-shaded uncertainty intervals Ω_l.

Typical repetitive controller design problem (5.14) is rendered tractable by application of the (generalized) KYP lemma and the resulting SDP is solved with [94]. Each of the problems (5.14) required to generate the results of this section is solved within 0.1 CPU second (Intel® Core™2 Duo T9300, 2.5 GHz, 3.5 GB of RAM). Generalized repetitive controller design problem (6.8), on the other hand, is handled with gridding, and SDPT3 [141, 149] requires about 30 CPU seconds to solve the resulting SOCP for $M = 299$, the highest value considered in this section (Intel® Core™2 Duo T9300, 2.5 GHz, 3.5 GB of RAM). The combination of the (generalized) KYP lemma and [94] is less appropriate for solving (6.8) as it invokes longer computational time (several CPU minutes) and requires more memory.

Modifying Sensitivity

This paragraph compares for $M = 144$ the optimal typical repetitive controller $K_{tRC,144}(z)$ and optimal generalized repetitive controller $K_{gRC,144}(z)$ that yield $\gamma_{np} = 1.3$, while $\delta = 1\%$ uncertainty on f_p is considered. Hence, $K_{tRC,144}(z)$ is obtained by solving (5.14) with $\mu = 2$, while $K_{gRC,144}(z)$ is obtained by solving (6.8) with $M = 144$. Instead of tuning α such that the resulting controller yields $\gamma_{np} = 1.3$, it is more appropriate to add the constraint $\gamma_{np} \leq 1.3$ to (5.14) and (6.8). If α is small enough, this constraint will be active.

For the same level of nonperiodic performance, $\gamma_{np} = 1.3$, the generalized repetitive controller yields significantly better periodic performance than the typical repetitive controller: $K_{gRC,144}(z)$ yields $\gamma_p = 0.23$, whereas $\gamma_p = 0.61$ for $K_{tRC,144}(z)$. Figure 6.9 compares the amplitude FRFs of $M_S(z)$ obtained by $K_{tRC,144}(z)$ and $K_{gRC,144}(z)$, where the shaded bands indicate the uncertainty intervals Ω_l (2.4) corresponding to $\delta = 1\%$. For $K_{tRC,144}(z)$, this figure reveals the characteristic periodic behavior of $M_S(\omega)$ in the pass-band of $Q(z)$, which is imposed by the typical repetitive controller structure. This FRF periodicity restricts the ability of $K_{tRC,144}(z)$ to optimize γ_p in two ways. First, $K_{tRC,144}(z)$ automatically accounts for all harmonics in the pass-band of $Q(z)$ and hereby wastes control effort at the even harmonics, not present in the input. $K_{gRC,144}(z)$, on the other hand, only improves the closed-loop performance around the harmonics $l \in \mathcal{L}$. Second, contrary to $K_{gRC,144}(z)$, $K_{tRC,144}(z)$ cannot account properly for the growing uncertainty intervals Ω_l around the harmonics. For $K_{gRC,144}(z)$, constraint (6.8c) is active for all harmonics $l \in \mathcal{L}$, that is: in each of the gray-shaded uncertainty intervals there is at least one ω value for which $|M_S(\omega)| = \gamma_p = 0.23$. For $K_{tRC,144}(z)$, on the other hand, constraint (5.14c) is only active for $l = 7$, the highest harmonic in $\mathcal{L}$, and hence, this controller is overly robust at the lower harmonic frequencies. Although not illustrated by the simulation example, in a similar way the periodicity of $M_S(\omega)$ impedes a typical repetitive controller to properly account for harmonic dependent weights W_l.

Trade-off Curves

To allow for a more systematic comparison of the performance achievable by a $K_{tRC,144}(z)$ and a $K_{gRC,144}(z)$, Figure 6.10 shows the corresponding trade-off curves between γ_p and γ_{np}, for $\delta = 0\%$ (a), $\delta = 1\%$ (b) and $\delta = 2\%$ (c). These trade-off curves are computed by solving (5.14) with $\mu = 2$, and (6.8) with $M = 144$ for various values of α: by increasing α they are traced from left to right. For a given level of periodic performance improvement γ_p, the curve indicates the minimal level of nonperiodic performance degradation γ_{np} that has to be tolerated, or, *vice versa*, for a fixed level of allowable nonperiodic performance degradation, the trade-off curve indicates the best periodic performance improvement that can be achieved.

Figure 6.10(a) reveals that, if no uncertainty on f_p is present ($\delta = 0\%$), $K_{gRC,144}(z)$ can improve the periodic performance index of $K_{tRC,144}(z)$ with up to 0.47, while yielding the same level of nonperiodic performance degradation. Thanks to a generalized repetitive controller's capability to deal more efficiently with uncertainty on f_p, the advantage of $K_{gRC,144}(z)$ over $K_{tRC,144}(z)$ is larger for higher δ: for $\delta = 2\%$ and the same γ_{np} level, $K_{gRC,144}(z)$ allows decreasing the γ_p value of $K_{tRC,144}(z)$ with 0.57. Figures 6.10(b) and 6.10(c) reveal that for $\delta > 0\%$, a $K_{gRC,144}(z)$ is able to achieve γ_p values not attainable by a $K_{tRC,144}(z)$. For instance, in Figure 6.10(c) where $\delta = 2\%$, the γ_p lower bound equals 0.013 for $K_{gRC,144}(z)$, while 0.35 for $K_{tRC,144}(z)$.

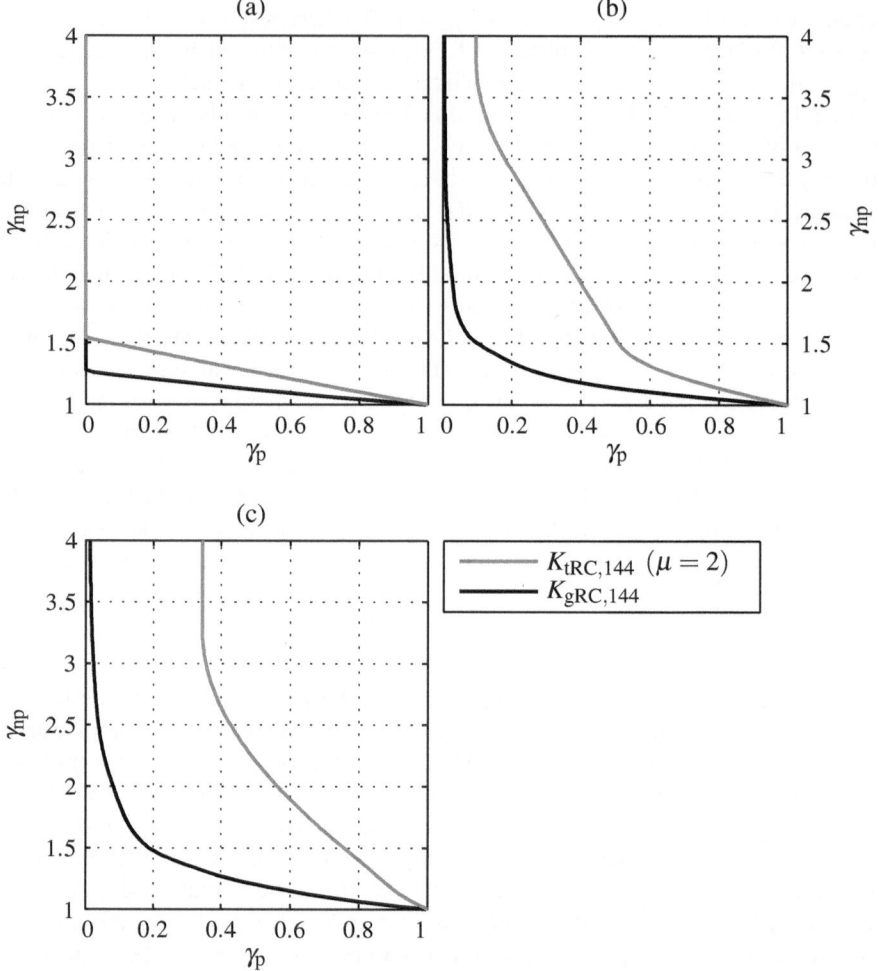

Fig. 6.10 Trade-off curves between γ_p and γ_{np} for $K_{tRC,144}$ and $K_{gRC,144}$: results for (a) $\delta = 0\%$; (b) $\delta = 1\%$; and (c) $\delta = 2\%$ uncertainty on f_p.

When is Typical Repetitive Control Optimal?

The structure of a typical repetitive controller dictates an FRF periodicity for $M_S(z)$ in the pass band of $Q(z)$, and hereby restricts the typical repetitive controller's ability to account for harmonic dependent weights W_l and uncertainty on the input period. On the other hand, typical repetitive controllers are expected to perform well in applications where these issues are not encountered, and the results presented in this paragraph confirm this expectation.

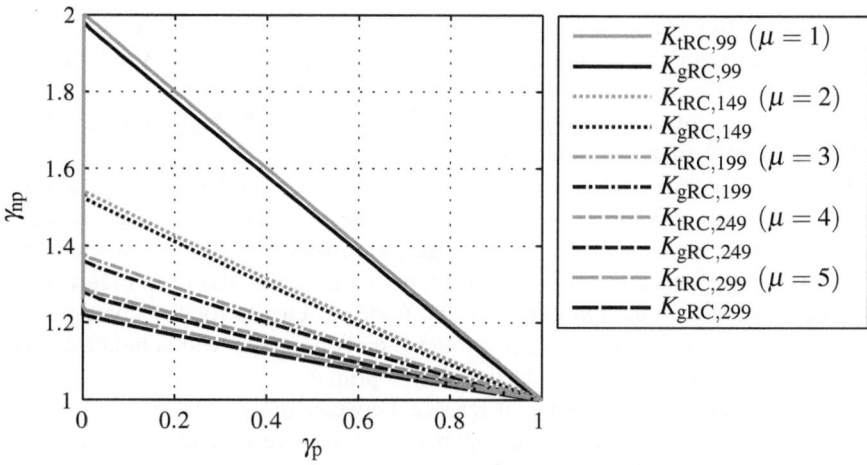

Fig. 6.11 Trade-off curves between γ_p and γ_{np} for various typical and generalized repetitive controllers. $\mathscr{L}$ is modified to $\mathscr{L} = \{0,1,2,3,4,5,6,7\}$, ω_{BW} is lowered to $2\pi173$ rad/s, and no uncertainty on f_p is considered: $\boldsymbol{\delta} = 0\%$.

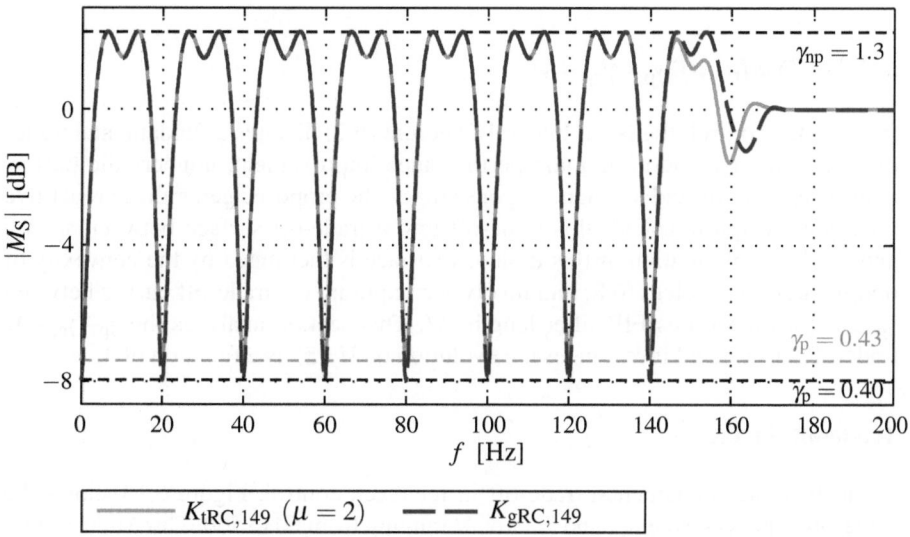

Fig. 6.12 Comparison of $|M_S(\omega)|$ for the $K_{tRC,149}(z)$ and $K_{gRC,149}(z)$ that yield $\gamma_{np} = 1.3$. $\mathscr{L}$ is modified to $\mathscr{L} = \{0,1,2,3,4,5,6,7\}$, ω_{BW} is lowered to $2\pi173$ rad/s, and no uncertainty on f_p is considered: $\boldsymbol{\delta} = 0\%$.

The simulation example is modified accordingly: all harmonics up to $l_{max} = 7$ are accounted for with equal weight: $W_l = 1$, for all l in $\mathscr{L} = \{0, 1, 2, 3, 4, 5, 6, 7\}$. In addition, $\delta = 0\%$, and ω_{BW} is lowered to $2\pi 173$ rad/s such that in order to meet its design specifications, $Q(z)$ needs to be of order $n_Q = 98$, the highest order allowed for a causal implementation (see Section 5.2.2).

Figure 6.11 compares the trade-off curves between γ_p and γ_{np} for typical repetitive controllers of various orders with the ones for the generalized repetitive controllers that yield the same transient response time for $M_S(z)$. For the typical repetitive controllers, these curves correspond to the results shown in Figure 5.6(a). For the modified simulation example, the trade-off curves for the corresponding typical and generalized repetitive controllers nearly coincide, which indicates that a typical repetitive controller is indeed close to optimal.

To investigate the remaining difference between the typical and generalized repetitive controllers, Figure 6.12 compares $|M_S(\omega)|$ corresponding to the $K_{tRC,149}(z)$ and $K_{gRC,149}(z)$ that yield $\gamma_{np} = 1.3$. Up to $l_{max}f_p = 140$ Hz, these FRFs are very similar, which confirms the good suboptimality of $K_{tRC,149}(z)$. The modifying sensitivities mainly differ between 140 Hz and 173 Hz. In this frequency range, $|M_S(\omega)|$ for $K_{tRC,149}(z)$ is determined by $Q(z)$, while $K_{gRC,149}(z)$ has more freedom to shape $|M_S(\omega)|$. This results in a minor performance improvement: $K_{gRC,149}(z)$ yields $\gamma_p = 0.40$ while $\gamma_p = 0.43$ for $K_{tRC,149}(z)$.

6.5.2   Trade-off $\gamma_p - \gamma_{np} - M$

In feedback control, the Bode Integral Theorem (6.7) dictates a fundamental trade-off between closed-loop periodic performance improvement, nonperiodic performance degradation and transient response time. The proposed generalized repetitive controller design translates this trade-off into a trade-off surface between γ_p, γ_{np} and M. The computation of this trade-off surface is facilitated by the convexity of optimal design problem (6.8) and involves computing the trade-off curves between γ_p and γ_{np} for various FIR filter lengths M. This section analyzes the $\gamma_p - \gamma_{np} - M$ trade-off surface and investigates the evolution of $M_S(z)$ as a function of M.

Trade-off Surface

To analyze the fundamental trade-off in feedback control, Figure 6.13 shows the trade-off curves between γ_p and γ_{np}, for M ranging from 50 to 500. Three uncertainty levels δ on the fundamental frequency are considered: $\delta = 0\%$ (a), $\delta = 1\%$ (b), and $\delta = 2\%$ (c). Each of the trade-off curves is computed by solving (6.8) for various weights α. Optimization problem (6.8) is rendered numerically tractable by gridding and the computational time required by SDPT3 [141, 149] to solve the resulting SOCP ranges from ± 5 CPU seconds for $M = 50$, to ± 500 CPU seconds for $M = 500$ (Intel® Core™2 Duo T9300, 2.5 GHz, 3.5 GB of RAM).

The trade-off curves shown in Figure 6.13 define the $\gamma_p - \gamma_{np} - M$ trade-off surface, which features a staircase-like behavior in M, since M is integer. The trade-off

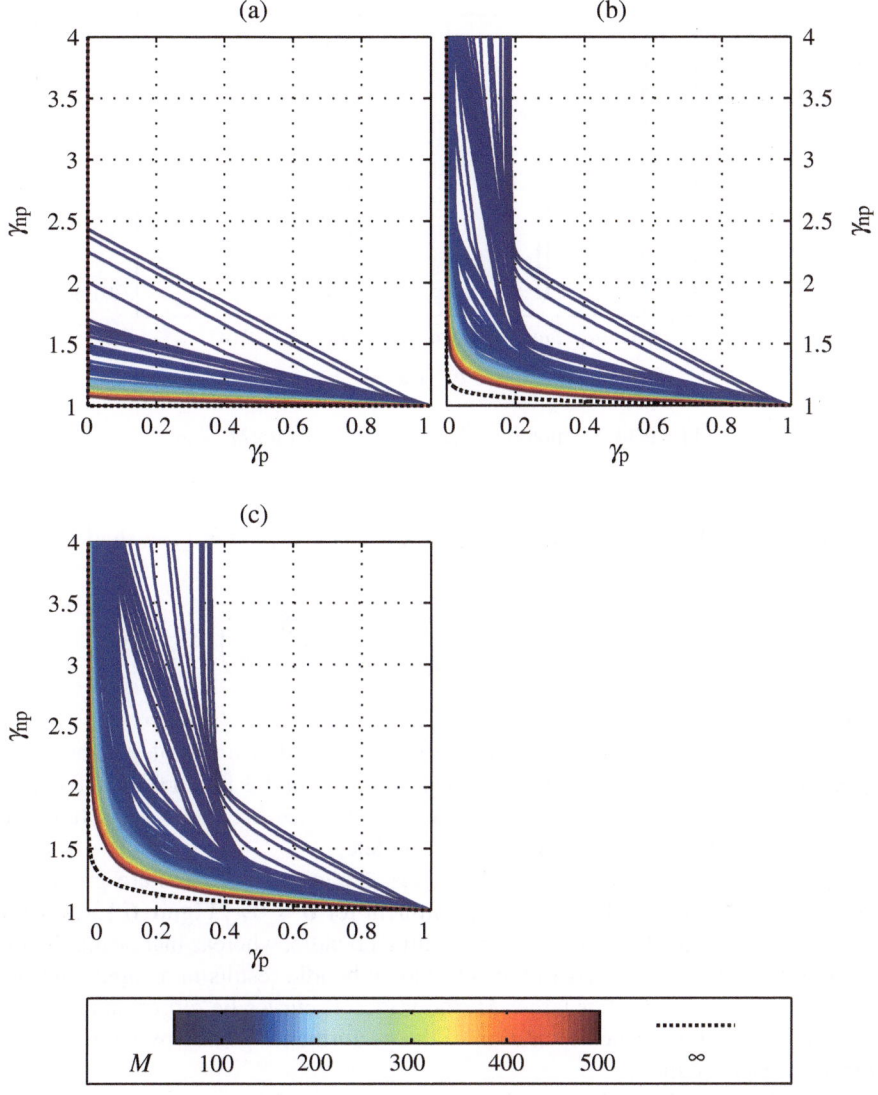

Fig. 6.13 Trade-off curves between γ_p and γ_{np} for M ranging from 50 to 500, while the black dotted line corresponds to the asymptotic curve for $M \to \infty$. Three uncertainty levels $\boldsymbol{\delta}$ on the fundamental frequency are considered: (a) $\boldsymbol{\delta} = 0\%$; (b) $\boldsymbol{\delta} = 1\%$; and (c) $\boldsymbol{\delta} = 2\%$.

curves mark the M stairs, where a generalized repetitive controller with given M can obtain all $\gamma_p - \gamma_{np}$ pairs above the corresponding trade-off curve.

The black dotted line in Figure 6.13 corresponds to the asymptotic trade-off curve between γ_p and γ_{np} for $M \to \infty$, and this curve is computed based on the asymptotic behavior of $|M_S(\omega)|$ for $M \to \infty$, shown in Figure 6.14. The Bode Integral

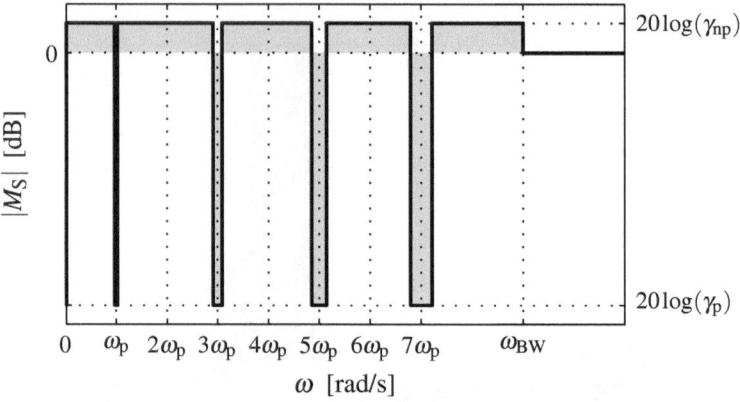

Fig. 6.14 Asymptotic behavior of $|M_S(\omega)|$ for $M \rightarrow \infty$.

Theorem (6.7) dictates equality of the positive and negative gray-shaded areas, which mathematically yields:

$$\gamma_{np} = \exp\left(-\ln(\gamma_p) \frac{\sum_{l \in \mathscr{L}} 2l\omega_p \boldsymbol{\delta}}{\omega_{BW} - \sum_{l \in \mathscr{L}} 2l\omega_p \boldsymbol{\delta}} \right). \tag{6.11}$$

If T_p is known with infinite accuracy, $\boldsymbol{\delta} = 0\%$, the trade-off between γ_p and γ_{np} vanishes for $M \rightarrow \infty$, as then, (6.11) yields $\gamma_{np} = 1$, independent of γ_p. The dotted lower bound for the trade-off curve between γ_p and γ_{np} holds for all add-on feedback controllers $K_{FB}(z)$, irrespective of their structure, since the set of basis functions (2.17) chosen for the Youla parameter is complete.

Figure 6.13 reveals that for low M values, the trade-off curves are grouped in bundles, where this bundling is most prominent for $\boldsymbol{\delta} = 0\%$, Figure 6.13(a). Little performance is gained by increasing M within a bundle, whereas increasing M such that the trade-off curve shifts to the subsequent bundle results in a significant performance improvement. For larger M, the trade-off curves lie closer and closer to each other, which indicates saturation of the performance gained by increasing the transient response time.

Figure 6.15 provides additional insight in the effect of the transient response time on the performance attainable by an add-on feedback controller. The black solid lines correspond to the cross-sections through the trade-off surfaces of Figure 6.13 at $\gamma_{np} = 1.3$, and consequently, these curves show the minimal γ_p value attainable by a generalized repetitive controller that yields $\gamma_{np} = 1.3$, as a function of M. The gray solid lines in Figure 6.15 correspond to optimal typical repetitive controllers, where their orders μ are translated into M by Equation 6.10. The black dashed lines indicate the asymptotic γ_p values for a generalized repetitive controller, predicted by (6.11), while the gray dashed lines indicate the asymptotic γ_p values for a typical repetitive controller, computed from (5.15).

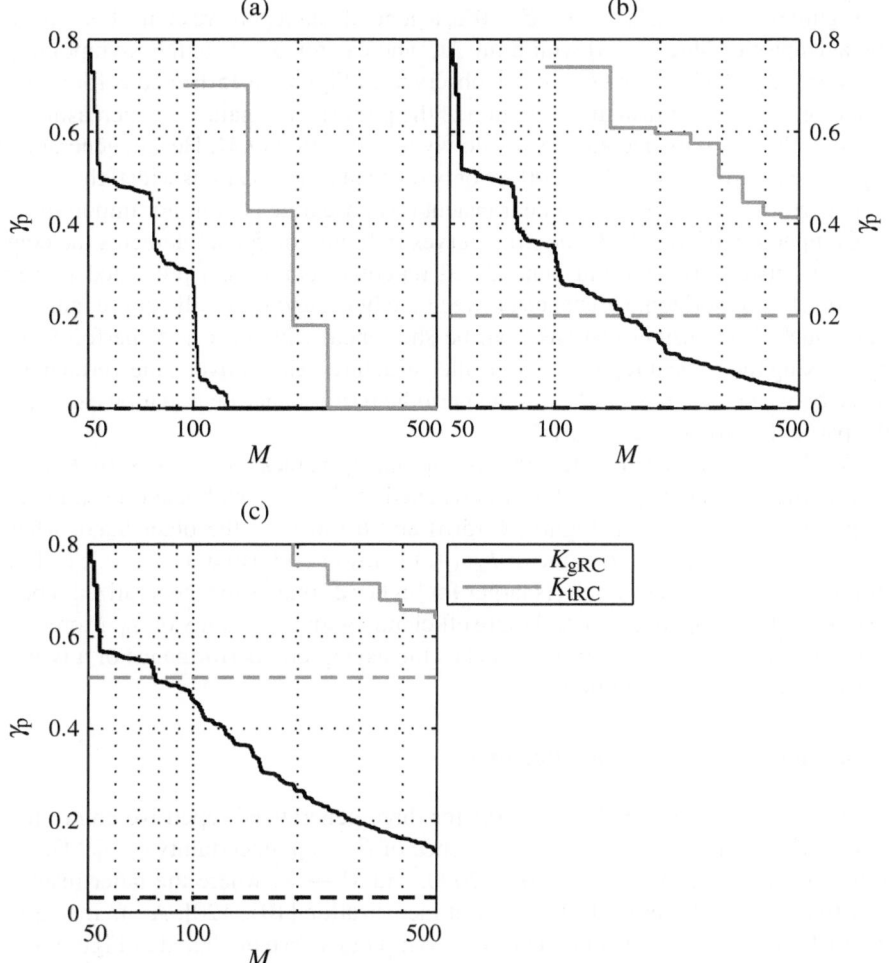

Fig. 6.15 Evolution of γ_p as a function of M (solid lines), for $\gamma_{np} = 1.3$ and three uncertainty levels δ on the fundamental frequency: (a) $\delta = 0\%$; (b) $\delta = 1\%$; and (c) $\delta = 2\%$. The black and gray lines respectively correspond to generalized and typical repetitive controllers, while the dashed lines indicate the corresponding asymptotic γ_p values for $M \to \infty$.

Since the trade-off curves in Figure 6.13 are grouped in bundles, the black solid lines in Figure 6.15 feature alternating parts with steep and shallow steps. These alternations are most prominent in Figure 6.15(a), where $\delta = 0\%$. From an engineering point of view, M values at the foot of a steep part are preferred: reducing M would result in a substantial loss of performance, whereas little performance is gained by increasing M.

Figure 6.15(a) reveals that for $\delta = 0\%$, a generalized repetitive controller reaches the asymptotic value $\gamma_p = 0$ with finite M, whereas for $\delta > 0\%$, the asymptotic γ_p value is only reached for $M \to \infty$, as observed in Figures 6.15(b) and 6.15(c). The latter figures also confirm the saturation of the performance gained by increasing M: for $M > 200$, γ_p only decreases very slowly as a function of M. Hence, generalized repetitive controllers with $M > 200$ are only relevant to applications with tight steady-state performance demands, while transient response time is of minor importance.

Comparison of the black and gray curves in Figure 6.15 complements the comparison between typical and generalized repetitive controllers presented in Section 6.5.1. The horizontal distance between the black and gray solid curve indicates how much the transient response can be shortened, without loss of performance, by relaxing the typical repetitive controller structure. Alternatively, the vertical distance between the curves indicates the periodic performance degradation caused by the particular structure of $K_{tRC}(z)$.

For $\delta = 0\%$, Figure 6.15(a), the asymptotic γ_p values for $M \to \infty$ of $K_{gRC}(z)$ and $K_{tRC}(z)$ coincide: $\gamma_p = 0$, but it is reached by $K_{gRC}(z)$ with a shorter transient response than $K_{tRC}(z)$. In Figures 6.15(b) and 6.15(c), on the other hand, where $\delta > 0\%$, the asymptotic γ_p value of $K_{gRC}(z)$ is lower compared to $K_{tRC}(z)$, and the difference between both values is larger for higher δ, thanks to a generalized repetitive controller's capability to deal more efficiently with uncertainty on f_p. Moreover, a generalized repetitive controller reaches the asymptotic performance of a typical repetitive controller with finite M.

Evolution of $M_S(z)$ as a Function of M

Figure 6.16 shows the FRFs of $M_S(z)$ for three generalized repetitive controllers, designed to yield $\gamma_p = 10^{-3}$ in the presence of $\delta = 1\%$ uncertainty on f_p. The results are shown for $M = 1000$, $M = 2000$, and $M \to \infty$, where the latter result is constructed from Figure 6.14 by spectral factorization [107, 120]. As M increases, the FRF of $M_S(z)$ indeed approaches the asymptotic curve predicted in Figure 6.14. However, approximating the sharp edges in the asymptotic $|M_S(\omega)|$ curve with a FIR parametrization gives rise to a Gibbs-like phenomenon [51], which is most prominent around 0 Hz. The results for the phase of $M_S(\omega)$ confirm that the asymptotic behavior of $M_S(z)$ around 0 Hz is particularly hard to catch with a FIR parametrization.

Figure 6.17(a) shows for the three considered generalized repetitive controllers the response of $M_S(z)$ to a sinusoidal input with frequency $3f_p = 60$ Hz and amplitude 1, while Figure 6.17(b) resumes the result for $M \to \infty$ in a larger time frame. The transient response of $M_S(z)$ is confined to M samples, and Figure 6.17(a) reveals that for smaller M, the transient response is faster: the response for $M \to \infty$ lags behind the response for $M = 2000$, which is slower than the response for $M = 1000$. For $M \to \infty$ the transient response consists of a decaying harmonic signal with frequency $3f_p = 60$ Hz, and its amplitude features a harmonic amplitude modulation of which the frequency evolves to $3f_p \delta = 0.6$ Hz. Figure 6.17(b) confirms that pushing a feedback controller design to the asymptotic $\gamma_p - \gamma_{np}$ trade-off curve, comes at the

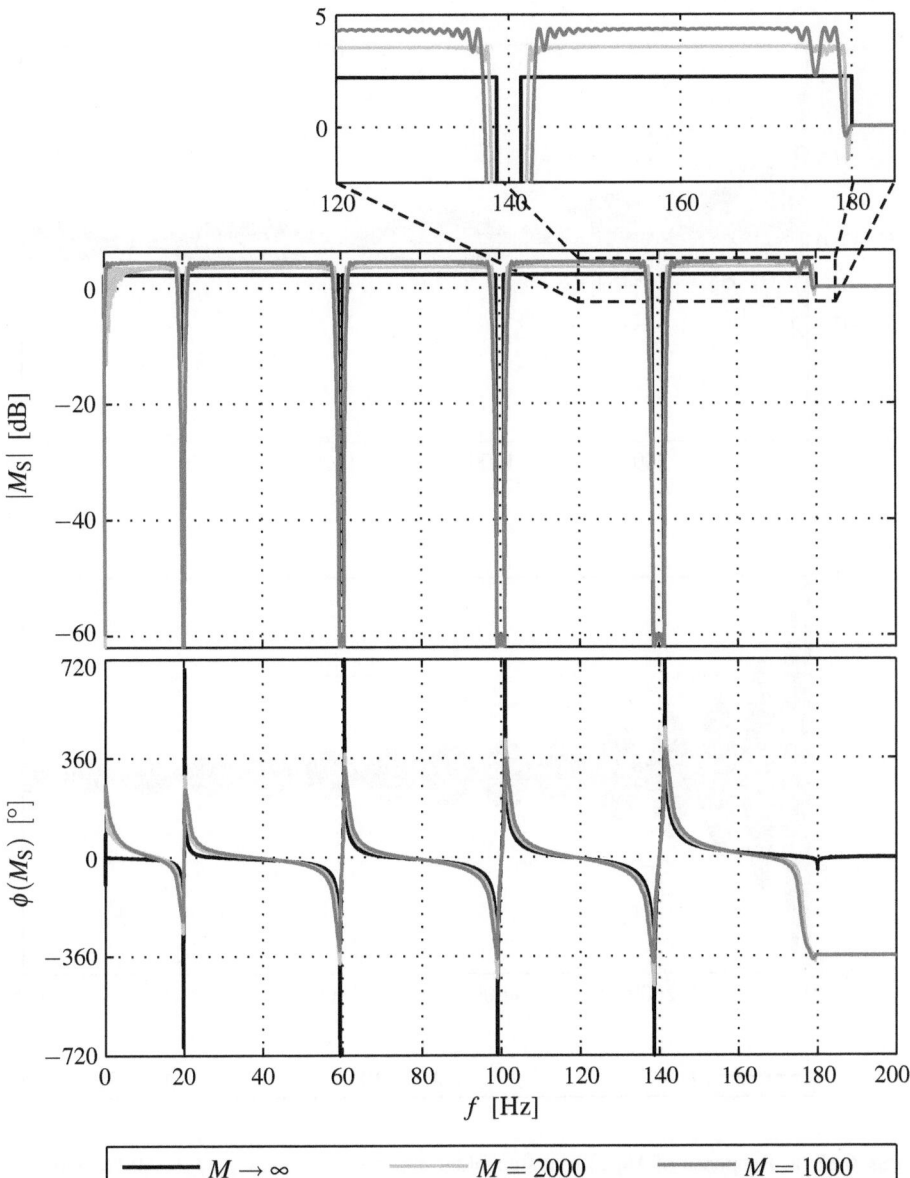

Fig. 6.16 FRF of $M_S(z)$ for three generalized repetitive controllers designed to yield $\gamma_p = 10^{-3}$ for $\delta = 1\%$.

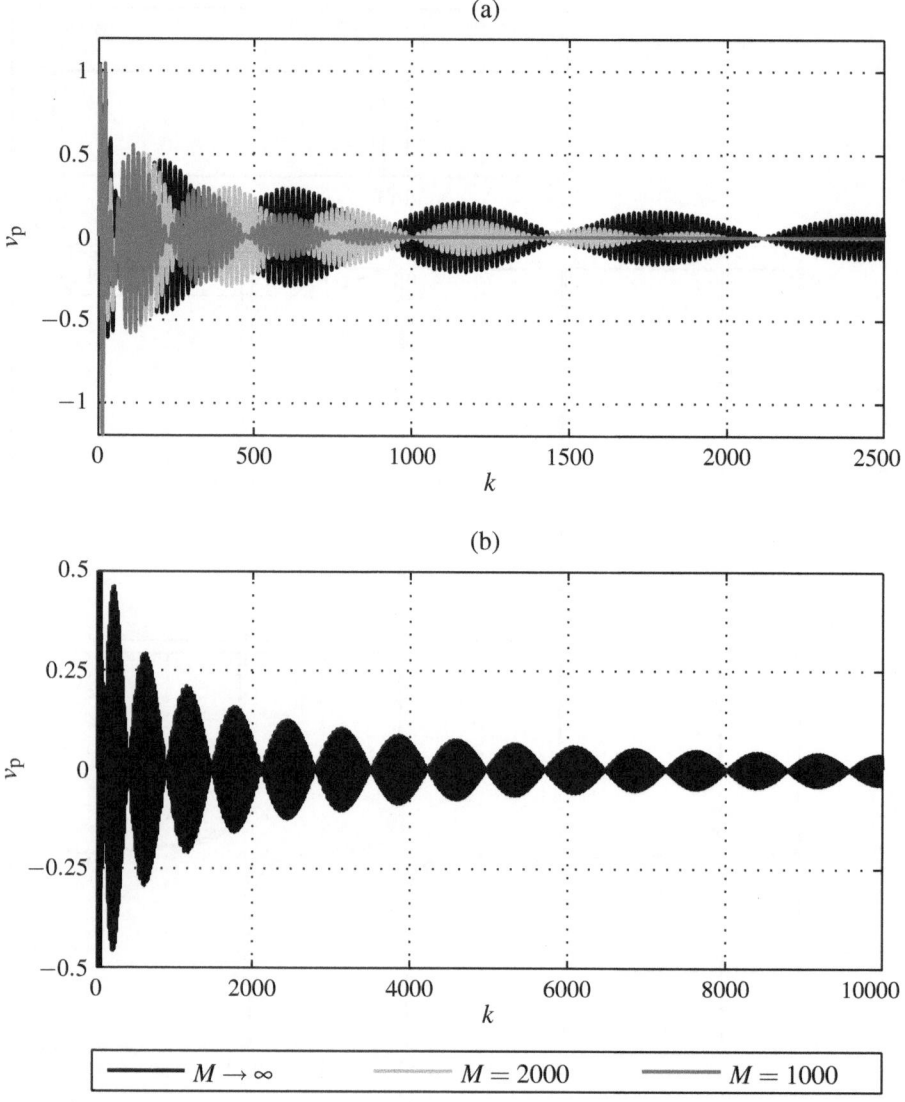

Fig. 6.17 (a) Response of $M_S(z)$ to a sinusoidal input with frequency $3f_p = 60$ Hz and amplitude 1 for three generalized repetitive controllers designed to yield $\gamma_p = 10^{-3}$ for $\boldsymbol{\delta} = 1\%$, while (b) resumes the result for $M \to \infty$ in a larger time frame.

price of a sluggish transient response: for $M \to \infty$ the transient response decays very slowly.

6.5.3 Comparison with the Literature

This section compares, for the considered simulation example, the developed generalized repetitive controller design with the current feedback controller designs of Hillerström and Sternby [65] and Köroğlu and Scherer [88]. For the sake of unified treatment, the latter design approaches are also applied to the add-on control configuration of Figure 6.1(a), and in addition, the following decomposition is enforced in all controllers:

$$K_{FB}(z) = \left[G(z)S_o(z)\right]_-^{-1} \overline{K}_{FB}(z) \, ,$$

such that

$$M_S(z) = \frac{1}{1 + G(z)S_o(z)K_{FB}(z)} = \frac{1}{1 + \left[G(z)S_o(z)\right]_+ \overline{K}_{FB}(z)} \, .$$

The approaches [65, 88] are applied to design $\overline{K}_{FB}(z)$ based on $M_S(z)$, hereby only accounting for $\left[G(z)S_o(z)\right]_+ = z^{-1}$, Equation 6.9. Since the design approaches of Hillerström and Sternby [65] and Köroğlu and Scherer [88] cannot cope with period-time uncertainty, $\delta = 0\%$ is applied throughout this section.

Hillerström and Sternby [65]

According to the Internal Model Principle, a feedback controller achieves perfect rejection/tracting of the periodic input $w_p(k)$ by placing closed-loop zeros at the harmonics $l \in \mathscr{L}$, that is: at the poles of $\Lambda(z)$, which lie on the unit circle. A controller $\overline{K}_{FB}(z)$ designed according to [65] places the zeros of $M_S(z)$ at these locations, while its poles are assigned to these locations, yet scaled with a factor $\beta < 1$. Figure 6.18 shows the resulting poles and zeros of $M_S(z)$ for $\beta = 0.995$ and $\beta = 0.95$.

Figure 6.19 evaluates these two controller designs in frequency and time domain: Figure 6.19(a) compares the resulting FRFs of $M_S(z)$, while Figure 6.19(b) shows the responses of $M_S(z)$ to a sinusoidal input with frequency $3f_p = 60$ Hz and amplitude 1. If β is close to one, the poles of $M_S(z)$ nearly cancel its zeros, which yields the benefit of a very small effect of these zeros on the closed-loop FRF: for $\beta = 0.995$, Figure 6.19(a) reveals sharp notches at the input harmonics, and hereby $\gamma_{np} = \|M_S(z)\|_\infty \approx 1$. On the other hand, this near pole-zero cancelation invokes a very sluggish transient response, as is clear from Figure 6.19(b). Placing the poles of $M_S(z)$ on the circle with radius β centered at the origin, results in a transient response that decays with β^k, and in Figure 6.19(b) the corresponding decay rates are indicated by dotted lines. Implied by the β^k decay, decreasing β from 0.995 to 0.95 results in much faster a transient response. On the other hand, Figure 6.19(b) reveals

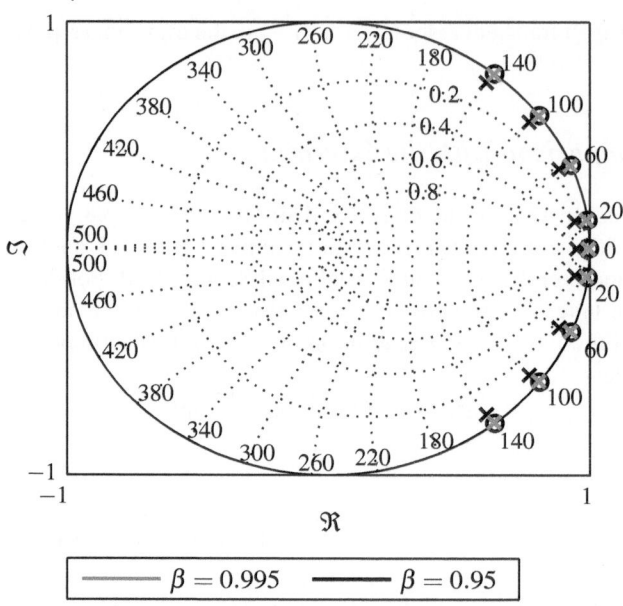

Fig. 6.18 Poles and zeros of $M_S(z)$ corresponding to two controllers $\overline{K}_{FB}(z)$ designed according to Hillerström and Sternby [65]. The grid indicates the natural frequency [Hz] and damping factor of the poles and zeros.

that with $\beta = 0.95$, the design of Hillerström and Sternby [65] is unacceptable for the considered simulation example: it improves the periodic performance by compromising the robust stability of the original feedback system. Preserving the original feedback system's robust stability requires $M_S(\omega) \approx 1$ from $\omega_{BW} = 2\pi180$ rad/s, whereas Hillerström and Sternby [65] provide no systematic way to include such constraint in the controller design.

Neither of the presented controllers designed according to Hillerström and Sternby [65] is compared to a generalized repetitive controller: with $\beta = 0.95$ the design of [65] is unacceptable, while the design of [65] with $\beta = 0.995$ is hard to approximate by a generalized repetitive controller, as was already observed in Figure 6.16.

Köroğlu and Scherer [88]

A controller $\overline{K}_{FB}(z)$ designed according to Köroğlu and Scherer [88] comprises two parts. One part guarantees good periodic performance while the complementary controller part handles the remaining design specifications by means of the Lyapunov shaping paradigm. Good periodic performance is translated into (6.5), and the two feedback controllers presented here differ in the adopted κ value: one enforces perfect periodic performance: $\kappa = 0$, while the other uses $\kappa = 0.5$.

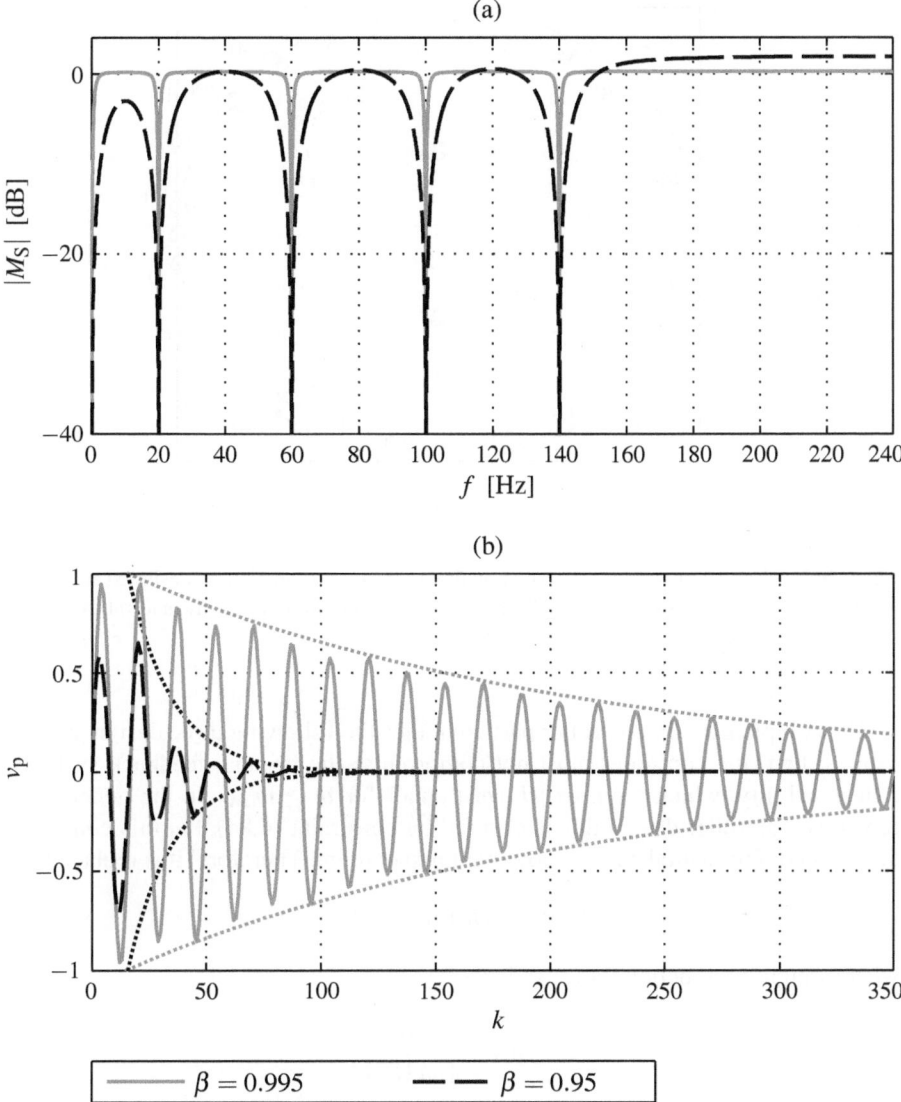

Fig. 6.19 Comparison of two controllers $\overline{K}_{FB}(z)$ designed according to Hillerström and Sternby [65]: (a) $|M_S(\omega)|$; and (b) response of $M_S(z)$ to a sinusoidal input with frequency $3f_p = 60$ Hz and amplitude 1. The dotted lines indicate the β^k decay rates of the transient responses.

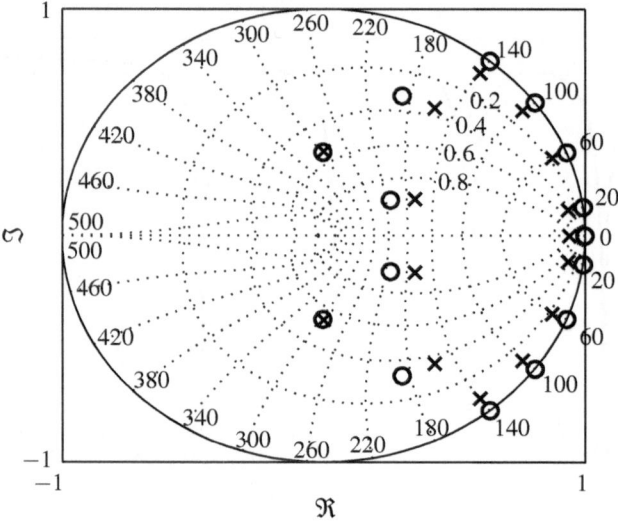

Fig. 6.20 Poles and zeros of $M_S(z)$ corresponding to the controller $\overline{K}_{FB}(z)$ designed according to Köroğlu and Scherer [88] with $\beta = 0.95$ and $\kappa = 0$. The grid indicates the natural frequency [Hz] and damping factor of the poles and zeros.

The complementary controller part combines the following three design specifications: first, the controller must not compromise the original feedback system's robust stability, which requires $|H_{rs}(\omega)|$ small for $\omega \geq \omega_{BW}$. In the generalized repetitive controller design this is ensured by constraint (6.8d), but this constraint does not comply with the Lyapunov shaping paradigm. Therefore, it is replaced by

$$\|W_{rs}(z)H_{rs}(z)\|_\infty \leq 1 \,,$$

where weighting function $W_{rs}(z)$ satisfies

$$W_{rs}(\omega) \approx \frac{\omega^4}{(2\pi 155)^4} \,,$$

to enforce a fourth-order roll-off of $|H_{rs}(\omega)|$ from 155 Hz. Second, the length of the closed-loop transient response should be restricted in the controller design, since otherwise, solutions similar to the result of Hillerström and Sternby [65] with $\beta = 0.995$ (Figure 6.19) are obtained. Similar to [65], the transient response is enforced to decay with at least 0.95^k by constraining the poles of $M_S(z)$ to the disc with radius $\beta = 0.95$ centered at the origin. Third, nonperiodic performance is optimized under the aforementioned constraints by minimizing $\gamma_{np} = \|M_S(z)\|_\infty$.

Figure 6.20 shows the poles and zeros of $M_S(z)$ corresponding to the feedback controller $\overline{K}_{FB}(z)$ designed according to the above described strategy with $\kappa = 0$. Figure 6.21 evaluates this controller design in frequency and time domain:

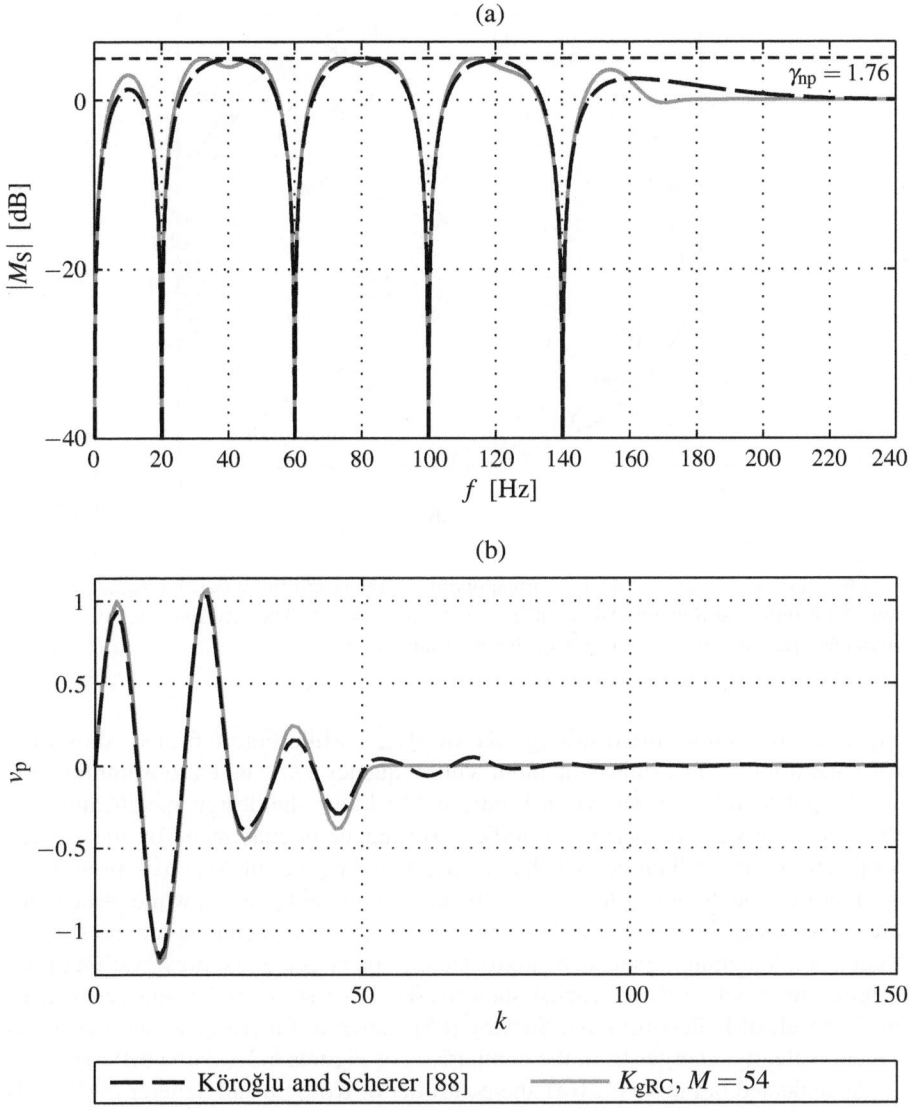

Fig. 6.21 Comparison of the controller $\overline{K}_{FB}(z)$ designed according to Köroğlu and Scherer [88] with $\beta = 0.95$ and $\kappa = 0$, and the corresponding generalized repetitive controller ($M = 54$): (a) $|M_S(\omega)|$; and (b) response of $M_S(z)$ to a sinusoidal input with frequency $3f_p = 60$ Hz and amplitude 1.

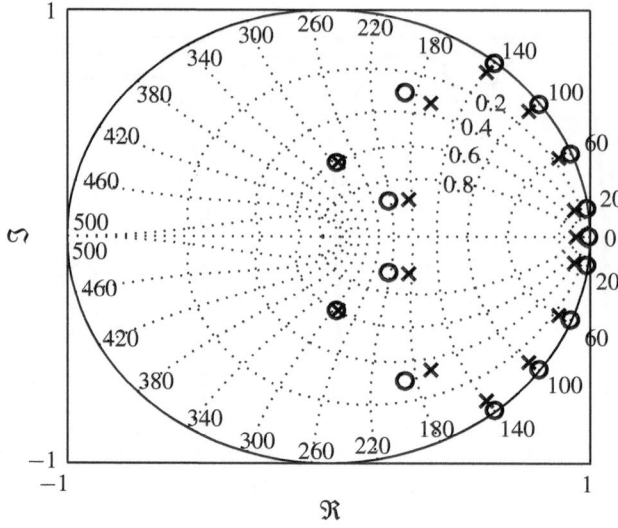

Fig. 6.22 Poles and zeros of $M_S(z)$ corresponding to the controller $\overline{K}_{FB}(z)$ designed according to Köroğlu and Scherer [88] with $\beta = 0.95$ and $\kappa = 0.5$. The grid indicates the natural frequency [Hz] and damping factor of the poles and zeros.

Figure 6.21(a) shows the resulting FRF of $M_S(z)$, while Figure 6.21(b) shows the response of $M_S(z)$ to a sinusoidal input with frequency $3f_p = 60$ Hz and amplitude 1.

The poles and zeros shown in Figure 6.20 address the design specifications in the following way: (i) perfect periodic performance is guaranteed by the closed-loop zeros on the unit circle, which coincide with the poles of $\Lambda(z)$; (ii) minimizing γ_{np} requires counteracting these zeros by nearly coinciding poles, while these poles are drawn away from the unit circle by the pole-placement constraint; and (iii) the three complementary resonance–antiresonance pairs guarantee preservation of the original feedback system's robust stability. This last set of poles and zeros lacks in the result of Hillerström and Sternby [65] shown in Figure 6.18, and hence, its function clearly emerges from the comparison of Figures 6.21(a) and 6.19(a).

As indicated in Figure 6.21(a), the controller of Köroğlu and Scherer [88] yields $\gamma_{np} = 1.76$, and Figure 6.21 compares this design with the fastest generalized repetitive controller, that is: the one with the smallest M, that satisfies $\gamma_{np} = 1.76$ and $\gamma_p < 10^{-6}$. This yields $M = 54$, and the obtained generalized repetitive controller resembles the controller designed by Köroğlu and Scherer [88], both in frequency and time domain. Figure 6.21(a) reveals the alternative implementation of the robust stability requirement, while Figure 6.21(b) shows that the generalized repetitive controller condenses the transient response of $M_S(z)$ into 54 samples.

In the second step, a controller $\overline{K}_{FB}(z)$ is designed according to Köroğlu and Scherer [88] with $\kappa = 0.5$, and this result yields $\gamma_{np} = 1.56$ and $\gamma_p = 0.18$.

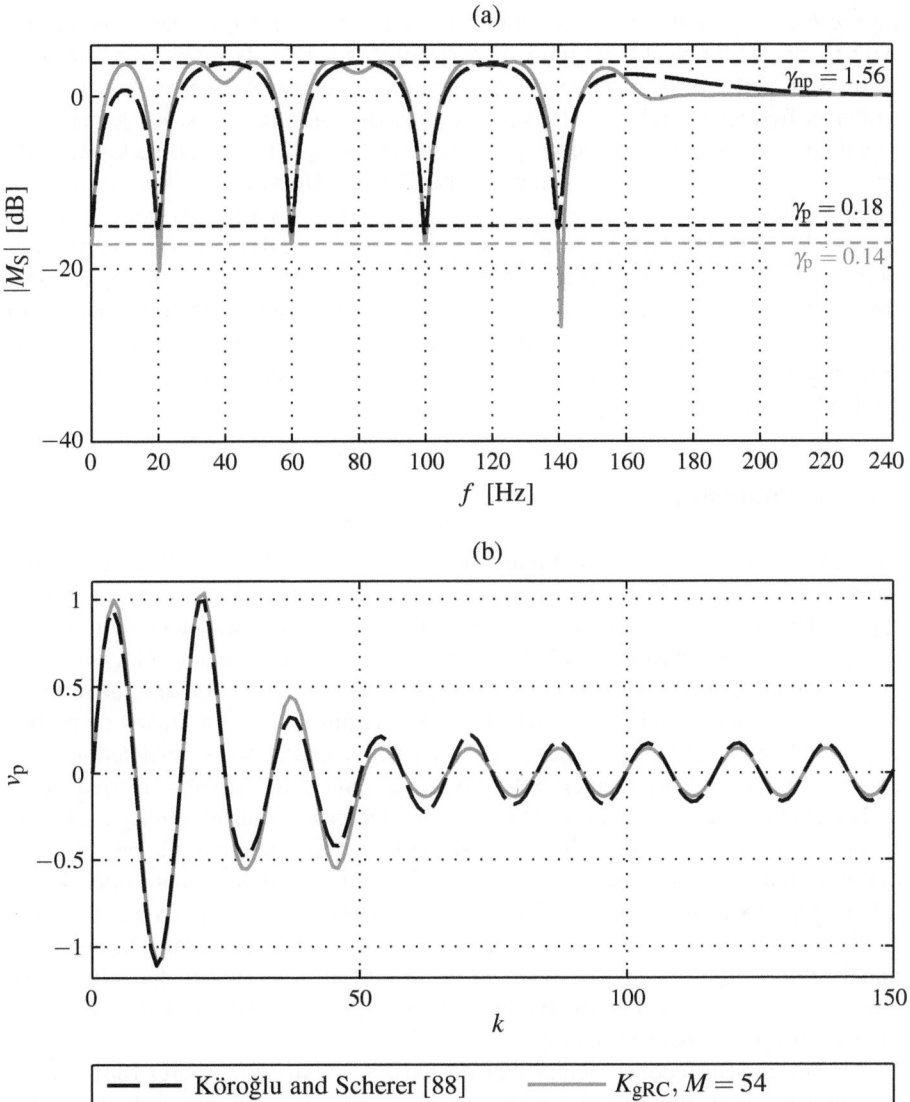

Fig. 6.23 Comparison of the controller $\overline{K}_{FB}(z)$ designed according to Köroğlu and Scherer [88] with $\beta = 0.95$ and $\kappa = 0.5$, and the corresponding generalized repetitive controller ($M = 54$): (a) $|M_S(\omega)|$; and (b) response of $M_S(z)$ to a sinusoidal input with frequency $3f_p = 60$ Hz and amplitude 1.

Figure 6.22 shows the corresponding poles and zeros of $M_S(z)$, and comparison with Figure 6.20 reveals that relaxing γ_p from 0 to 0.18 only involves a minor inward shift of the zeros of $M_S(z)$ on the unit circle. Figure 6.23 evaluates the controller in frequency and time domain and compares it with the fastest generalized repetitive controller that achieves $\gamma_{np} = 1.56$ and $\gamma_p \leq 0.18$, which yields $M = 54$ and $\gamma_p = 0.14$. Figure 6.23(a) compares the FRFs of $M_S(z)$ for the two controllers, while Figure 6.23(b) shows the responses of $M_S(z)$ to a sinusoidal input with frequency $3f_p = 60$ Hz and amplitude 1. Figure 6.23(a) reveals a different behavior of both controllers between 0 Hz and 20 Hz, in addition to the alternative implementation of the robust stability requirement. Figure 6.23(b) shows a slightly more gruff transient response for the generalized repetitive controller, due to the finite impulse response of $M_S(z)$. In addition, the small difference between the γ_p values of the two controllers prevails in steady state.

6.6 Conclusion

This chapter applies the methodology of Chapter 2 to design a feedback controller for a discrete-time SISO LTI system facing periodic inputs. This results in a novel type of feedback controllers, called generalized repetitive controllers, that encompasses typical, one-period delay based repetitive controllers as a special case. The generalized repetitive controller design allows a systematic and quantitative treatment of combined nonperiodic and period-uncertain inputs: by means of performance indices γ_p and γ_{np}, the generalized repetitive controller is designed to yield an optimal trade-off between closed-loop periodic and nonperiodic performance.

Dictated by the Bode Integral Theorem, any feedback controller design is bound to a trade-off between closed-loop periodic performance, nonperiodic performance and transient response time. The presented generalized repetitive controller design allows a quantitative analysis of this trade-off, by translating it into a trade-off surface between periodic performance index γ_p, nonperiodic performance index γ_{np} and the finite impulse response length of the (modifying) sensitivity. In addition, the performance loss caused by the typical repetitive controller structure is analyzed in view of this performance trade-off.

Among current feedback controller designs, Köroğlu and Scherer [88] is most related to the presented generalized repetitive controller design, where the presented design approach is shown able to generate controllers similar to the ones obtained by Köroğlu and Scherer [88]. On the other hand, the systematic treatment of period-time uncertainty that is presented in this chapter, is innovative with respect to the literature.

Chapter 7
Experimental Validation on an Active Air Bearing Setup

7.1 Introduction

In repetitive control, the Bode Integral Theorem [10, 21, 22, 49, 69, 138], dictates a fundamental trade-off between improved closed-loop periodic performance, degraded nonperiodic performance and transient response time. Relying on the design methodology of Chapter 2, Chapter 5 develops an optimal repetitive controller design, where optimality is translated into an optimal trade-off between performance indices γ_p and γ_{np}, which respectively quantify the closed-loop periodic and nonperiodic performance. Index γ_p explicitly accounts for period-time uncertainty, while transient response time, the third issue involved in the performance trade-off is determined by the repetitive controller order μ. For a given order μ, the remaining trade-off between periodic and nonperiodic performance is translated into a trade-off curve between γ_p and γ_{np} (see Section 5.4.1).

The purpose of this chapter is twofold. First, the value of the trade-off curves between γ_p and γ_{np} in dealing with the performance trade-off in repetitive control is experimentally demonstrated on an active air bearing setup [110]. The control objective is to reduce the error motion of the spindle's axis of rotation by appropriate actuation of an active journal bearing. This error motion, being due to mass unbalance and profile errors of the bearing parts, is periodic with the spindle rotation, leaving measurement noise as the sole nonperiodic input to the control problem. Comparison of various Pareto optimal repetitive controller designs reveals that superior reduction of the periodic error motion comes at too high a price of measurement noise amplification, which deteriorates the overall closed-loop performance. This way, the experimental results sustain the practical relevance of performance indices γ_p and γ_{np}, as well as the corresponding trade-off curves.

Second, the relation is investigated between the steady-state performance indices γ_p and γ_{np}, and the adaptive performance of the repetitive controller during large variations of the spindle's rotational speed setpoint, where the adaptive implementation is adopted from [42, 139]. The experiments indicate that good robust periodic performance for period-time uncertainty translates into good adaptive performance

G. Pipeleers et al.: Optimal Linear Controller Design for Periodic Inputs, LNCIS 394, pp. 119–140.
springerlink.com

during small variations of the spindle's rotational speed. Large variations, on the other hand, are no longer encompassed in the period-time uncertainty and cause performance degradation.

As periodic disturbances are characteristic for spindle applications, repetitive control is not new in the field of active bearing control. In active air bearing applications, repetitive control has been applied to overcome the low stiffness and damping of the air film [5, 7, 67, 68], and in some magnetic bearing systems, repetitive control has been used to increase the rotational accuracy [161, 162]. While in these applications the repetitive controller is restricted to a basic, first-order design, more advanced repetitive controller designs have been experimentally validated for disk drive servo systems [135, 136], a major industrial application of repetitive control. However, all these repetitive controllers enforce perfect suppression of the periodic disturbances without investigating the consequences of the performance trade-off in repetitive control on the overall performance.

This chapter is organized as follows. Section 7.2 describes the experimental setup and the corresponding control configuration. Section 7.3 elaborates on the repetitive controller design, while Section 7.4 experimentally validates the selected controllers. Section 7.5 concludes the chapter.

7.2 Experimental Setup

The experimental setup comprises the active air bearing prototype presented in [6] and is described in Section 7.2.1. Section 7.2.2 details the corresponding control configuration, while Section 7.2.3 discusses the parametric identification of the test setup.

7.2.1 Description

The experimental setup is depicted in Figure 7.1, where Figure 7.1(a) shows a top view on the front part of the setup and Figure 7.1(b) illustrates the active air bearing layout.

At the rear end of the setup an asynchronous motor drives the spindle, where the rotational speed is controlled in open loop by a frequency converter. The spindle is supported by aerostatic bearings: the axial (thrust) and rear radial (journal) bearings are passive whereas the front radial bearing is active. The active bearing comprises a compliant bearing surface composed of four lands, which are each supported on a row of two piezo-actuators. Those actuators deform the bearing surface in a controlled manner so as to induce a radial force on the shaft via the air film. In the case of a radial spindle bearing, the shaft has two degrees of freedom normal to its axis. Consequently, two displacement sensors are needed, one for each of these two principal directions. Capacitive sensors are employed (Lion Precision) and a reference

(a)

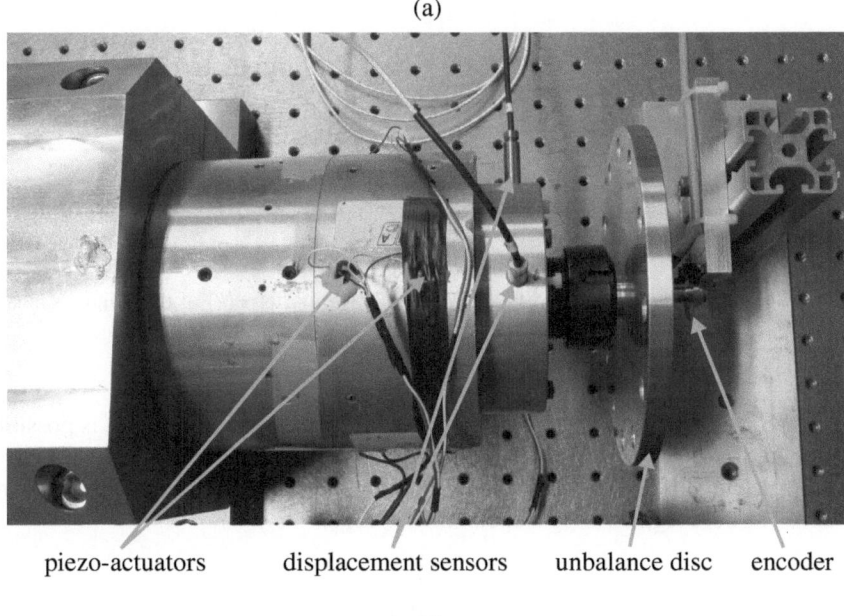

piezo-actuators   displacement sensors   unbalance disc   encoder

(b)

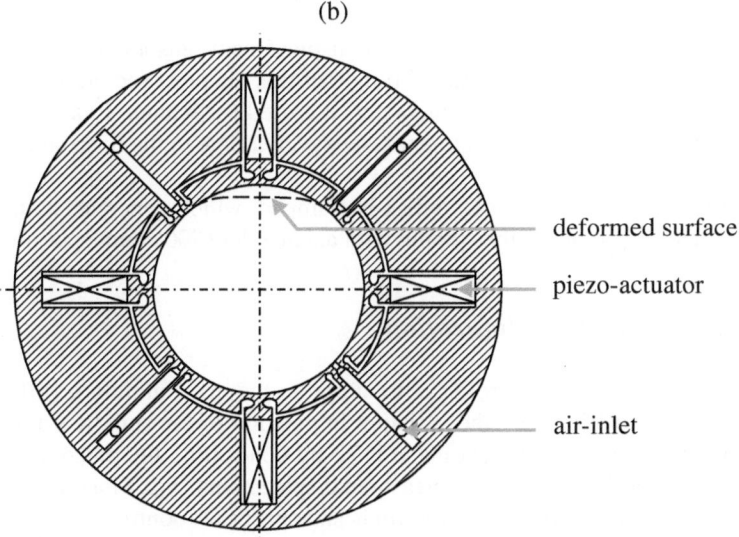

deformed surface

piezo-actuator

air-inlet

Fig. 7.1 Active air bearing setup: (a) top view on the front part of the experimental setup; and (b) illustration of the active air bearing layout (not to scale), where the dashed line illustrates the bearing surface deformation (not to scale) due to actuation of the top piezo-actuator.

target ring is machined in situ on the spindle nose. The two measurement directions are aligned horizontally and vertically, and coincide with the working directions of the piezo-actuators. When driving the diametrically opposite piezo-actuators in

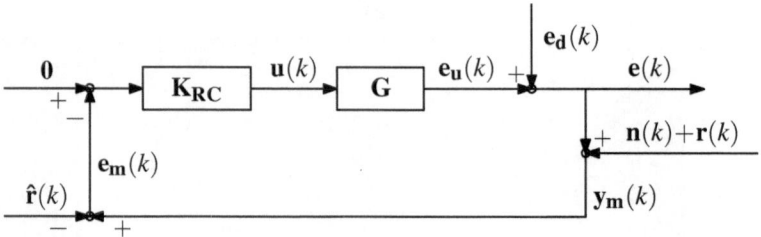

Fig. 7.2 Control configuration used to suppress the error motion $\mathbf{e}(k)$ of the spindle's axis of rotation by repetitive control.

anti-phase, both "pushing" and "pulling" in the two principal directions is possible. As a result, only two scalar control inputs, denoted u_h [V] and u_v [V], are needed to determine the amplifier inputs for the horizontal and vertical piezo-actuators, respectively. These signals are grouped into the vector

$$\mathbf{u} = \begin{bmatrix} u_h \\ u_v \end{bmatrix},$$

and in the same way, the outputs of the displacement sensors, denoted $y_{m,h}$ [m] and $y_{m,v}$ [m], are grouped into $\mathbf{y_m}$. Throughout this chapter, vectors and matrices are indicated in bold, while plain characters are used for scalars. In a two-dimensional vector, the first and second element respectively relate to the horizontal and vertical direction.

At the nose of the spindle, a disk is clamped, which generates a 46 g mm mass unbalance and connects the spindle with an encoder (500 counts per rotation).

7.2.2 Control Configuration

The objective of the controller is to suppress the error motion $\mathbf{e}$ [m] of the spindle's axis of rotation, i.e., the variation in position of the spindle's axis of rotation observed at the measurement plane [17], by controlling the piezo-actuators of the active air bearing based on the measured displacements $\mathbf{y_m}$. Since $\mathbf{e}$ is periodic with the spindle rotation, repetitive control is an appropriate control strategy. The control problem is handled in discrete time, where the sample frequency f_s equals 10 kHz and the corresponding control configuration is shown in Figure 7.2. Mainly motivated by practical issues, the experimental validation of the repetitive controllers is confined to spindle speeds between 900 rpm and 1200 rpm.

Error motion $\mathbf{e}(k)$ is the sum of two contributions: (i) $\mathbf{e_d}(k)$ [m] related to the mass unbalance and profile errors of the bearing parts; and (ii) $\mathbf{e_u}(k)$ [m] caused by actuation of the active air bearing. The displacement sensors generate the measurement $\mathbf{y_m}(k)$, which corresponds to the actual error motion $\mathbf{e}(k)$ supplemented with (i) (stochastic) measurement noise $\mathbf{n}(k)$ [m]; and (ii) a (periodic) systematic error $\mathbf{r}(k)$ [m] due to the roundness error of the reference target ring for the displacement sensors.

As suppressing the error motion $\mathbf{e}(k)$ constitutes the control objective, the repetitive controller $\mathbf{K_{RC}}(z)$ should not respond to the systematic measurement error $\mathbf{r}(k)$. To this end, the measured error motion $\mathbf{e_m}(k)$ [m] is constructed from measurements $\mathbf{y_m}(k)$ by subtracting an estimate $\hat{\mathbf{r}}(k)$ [m] of the roundness error $\mathbf{r}(k)$. To compute $\hat{\mathbf{r}}(k)$ at each time instant k, the roundness error is determined *a priori* as a function of the spindle angle θ [rad] using the method of master reversal proposed by Donaldson [41]. During rotation, $\theta(k)$ is measured by the encoder, which yields $\hat{\mathbf{r}}(k)$ as a function of time.

In view of the add-on repetitive control setup adopted in Chapter 5, the so-called original feedback controller $\mathbf{K_o}(z)$ is omitted: $\mathbf{K_o}(z) = \mathbf{0}$. This choice is allowed as the experimental setup is open-loop stable. Moreover, it yields the best nonperiodic performance since measurement noise $\mathbf{n}(k)$ is the sole nonperiodic input to the control problem.

7.2.3 Identification

For the considered control problem, the plant $\mathbf{G}(z)$ corresponds to the two-by-two system with input $\mathbf{u}(k)$ and output $\mathbf{e_u}(k)$:

$$\underbrace{\begin{bmatrix} e_{\mathbf{u},h}(k) \\ e_{\mathbf{u},v}(k) \end{bmatrix}}_{\mathbf{e_u}(k)} = \underbrace{\begin{bmatrix} G_h(q) & G_{hv}(q) \\ G_{vh}(q) & G_v(q) \end{bmatrix}}_{\mathbf{G}(q)} \underbrace{\begin{bmatrix} u_h(k) \\ u_v(k) \end{bmatrix}}_{\mathbf{u}(k)}.$$

This system is identified in open loop with separate excitation of the horizontal and vertical piezo-actuators, and the identification is performed at three spindle speeds: 900 rpm, 1050 rpm and 1200 rpm. The system is excited between 1 Hz and 2500 Hz, where the fundament of the excitation signals is a random-phase multi-sine [109] with a frequency resolution of 1 Hz and flat amplitude spectrum. Excitation frequencies coinciding with harmonics of the spindle's rotational speed are eliminated to prevent that the error motion $\mathbf{e_d}(k)$ (due to mass unbalance and profile errors of the bearing parts) hampers the identification.

In the first step, nonparametric FRF estimates are obtained for the four plant components and at the three considered spindle speeds. Figure 7.3 shows these nonparametric FRFs, and as they only reveal a limited effect of the spindle's rotational speed on the plant dynamics (between 900 and 1200 rpm), all results are shown in gray (three spindle speeds and various experiments at each speed). Figure 7.3 reveals that up to 600 Hz the off-diagonal gains $|G_{hv}(\omega)|$ and $|G_{vh}(\omega)|$ are on average 20 dB smaller than the diagonal gains $|G_h(\omega)|$ and $|G_v(\omega)|$. To further analyze the interaction between the horizontal and vertical control direction, Figure 7.4 shows the interaction measure μ proposed by [53], as a function of frequency:

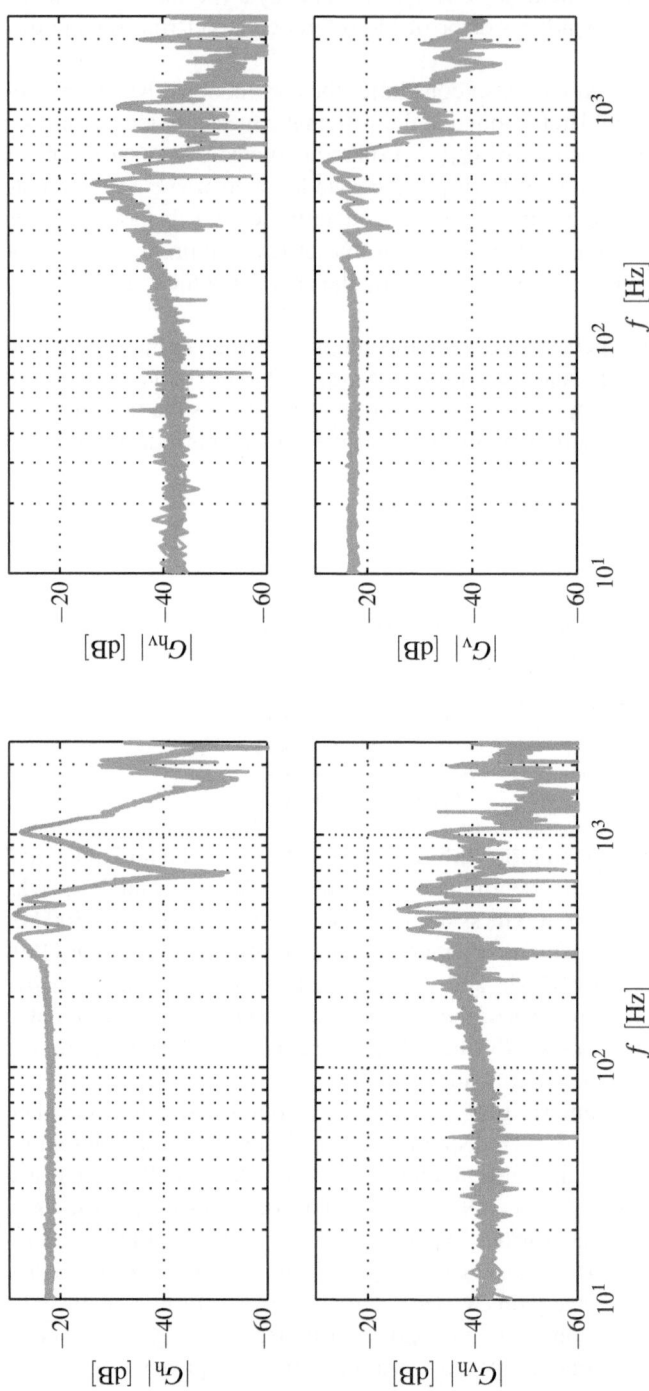

Fig. 7.3 Amplitudes of all the nonparametric FRF estimates obtained for the four plant components (three spindle speeds and various experiments at each speed).

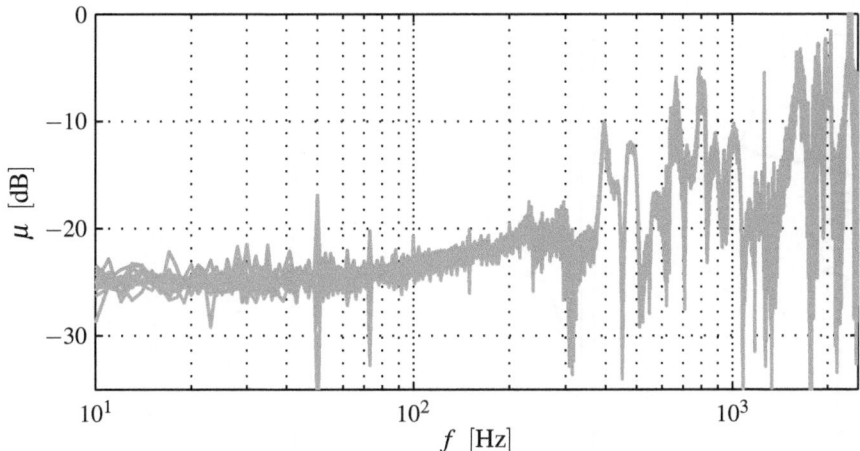

Fig. 7.4 Interaction measure μ as a function of frequency, computed from the nonparametric FRF estimates shown in Figure 7.3.

$$\mu(\omega) = \sqrt{\left| \frac{G_{hv}(\omega)G_{vh}(\omega)}{G_h(\omega)G_v(\omega)} \right|} \, .$$

This figure confirms the dominance of the diagonal plant components up to 600 Hz. Around 400 Hz, μ increases to -10 dB due to the combination of an antiresonance in $G_h(z)$ and an increase of the off-diagonal gains (see Figure 7.3).

In the second step, a parametric model $\widehat{\mathbf{G}}(z)$ is identified for $\mathbf{G}(z)$ based on the obtained nonparametric FRF estimates. This model should be accurate up to 600 Hz, since extensive simulations reveal that the active frequency range of the controller is restricted to this frequency by the uncertainty on the antiresonance of $G_h(z)$ around 680 Hz. Based on the dominance of the diagonal plant components, a decoupled plant model is proposed

$$\widehat{\mathbf{G}}(z) = \begin{bmatrix} \widehat{G}_h(z) & 0 \\ 0 & \widehat{G}_v(z) \end{bmatrix} , \tag{7.1}$$

where the SISO models $\widehat{G}_h(z)$ (9th-order) and $\widehat{G}_v(z)$ (17th-order) are identified using the `fdident` toolbox [85]. Figure 7.5 evaluates these models by comparing their FRFs with the nonparametric FRF estimates for the diagonal plant components. Implied by the use of model $\widehat{\mathbf{G}}(z)$, the plant is assumed decoupled for the controller design and the simulations.

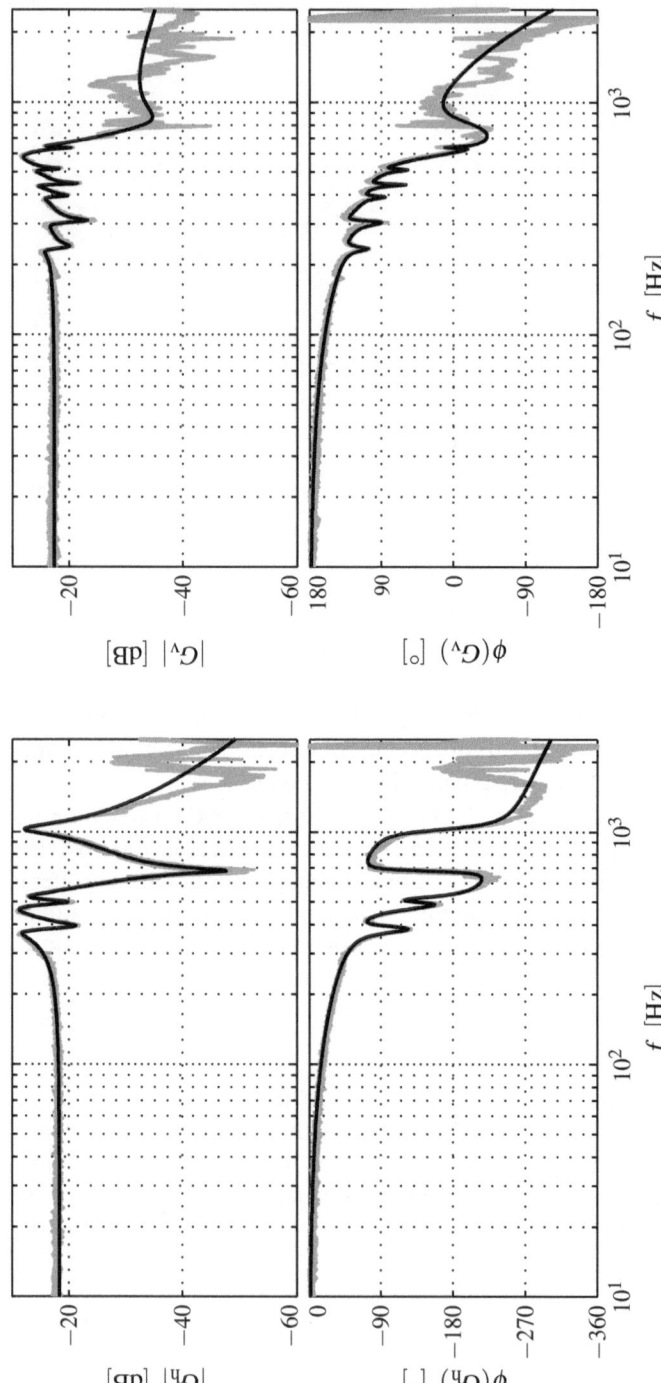

Fig. 7.5 Comparison of the FRFs of the parametric models $\widehat{G}_h(z)$ and $\widehat{G}_v(z)$ (black) with the corresponding nonparametric FRF estimates (gray).

7.3 Repetitive Controller Design

This section elaborates on the design of repetitive controller $\mathbf{K_{RC}}(z)$. Relying on the decoupled plant model $\widehat{\mathbf{G}}(z)$, a decoupled design of $\mathbf{K_{RC}}(z)$ suffices:

$$\mathbf{K_{RC}}(z) = \begin{bmatrix} K_{RC,h}(z) & 0 \\ 0 & K_{RC,v}(z) \end{bmatrix} . \tag{7.2}$$

The SISO repetitive controllers $K_{RC,h}(z)$ and $K_{RC,v}(z)$ are designed according to Chapter 5. Section 7.3.1 briefly summarizes the repetitive controller design, while Section 7.3.2 discusses the selected repetitive controllers for the control problem at hand.

Generalized repetitive controllers designed according to Chapter 6 are also experimentally validated, but these results are not presented here. The generalized repetitive controllers only achieve a limited performance improvement compared to the (typical) repetitive controllers, which is related to the large number of harmonics to be suppressed, and all harmonics except $l = 1$ having a similar contribution to $\mathbf{e_d}(k)$ (see Figure 7.10(a)). In addition, (typical) repetitive controllers have the advantage of facile adaptive implementation for varying spindle speeds (see Section 7.4.3).

Concerning the rotational speed of the spindle, the following notation is used: the desired rotational speed used as input for the frequency converter that drives the asynchronous motor, is indicated by $f_{p,des}$ [Hz]. However, a repetitive controller requires the period to contain an integer number N of samples. Hence, assuming that f_s cannot be changed, the best the repetitive controller can do is to account for $f_p = f_s/N$ [Hz], where

$$N = \text{int}(f_s/f_{p,des}) , \tag{7.3}$$

and $\omega_p = 2\pi f_p$ [rad/s]. The actual rotational speed of the spindle is denoted by $f_{p,\delta}$ [Hz], $\omega_{p,\delta} = 2\pi f_{p,\delta}$ [rad/s], and may deviate from both $f_{p,des}$ and f_p, where δ corresponds to the relative deviation from f_p and is bounded by $\boldsymbol{\delta}$ (2.3).

7.3.1 Repetitive Controller Design Strategy

The SISO repetitive controllers $K_{RC,h}(z)$ and $K_{RC,v}(z)$ feature the structure of Figure 7.6, where only the design of the filter $L(z)$ depends on the control direction. The horizontal repetitive controller equals

$$K_{RC,h}(z) = L_h(z) \frac{\chi(z)Q(z)}{1 - \chi(z)Q(z)} , \tag{7.4}$$

where

$$\chi(z) = \sum_{m=1}^{\mu} \chi_m z^{-mN} , \tag{7.5}$$

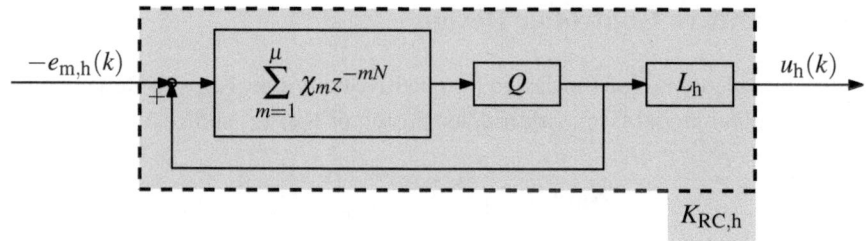

Fig. 7.6 Structure of the horizontal SISO repetitive controller, where the vertical equivalent is obtained by replacing subscript $(\cdot)_h$ by $(\cdot)_v$.

and μ is called the order of the repetitive controller. Under the assumption of a decoupled plant, the decoupled controller $\mathbf{K_{RC}}(z)$ (7.2) gives rise to a decoupled closed-loop sensitivity function, of which the horizontal diagonal element is given by

$$S_h(z) = \frac{1 - \chi(z)Q(z)}{1 - \chi(z)Q(z)\left[1 - L_h(z)G_h(z)\right]} . \qquad (7.6)$$

The vertical equivalents of (7.4) and (7.6) are obtained by replacing the subscript $(\cdot)_h$ by $(\cdot)_v$.

Filters $Q(z)$ and $L(z)$ guarantee robust stability of the closed-loop system, whereas the closed-loop performance is optimized through the design of parameters χ_m.

Stability

To achieve robust closed-loop stability, the filters $L(z)$ and $Q(z)$ are designed in accordance with the common procedure in repetitive control (see Section 5.2.2):

- $L(z)$ is designed as the series connection of the inverse plant model and a low-pass filter $F(z)$:

$$L_h(z) = \widehat{G}_h(z)^{-1}F(z), \qquad L_v(z) = \widehat{G}_v(z)^{-1}F(z) .$$

  Filter $F(z)$ is designed as a low-pass zero-phase FIR filter of order 300 with cut-off frequency 640 Hz, and it is added to improve robust closed-loop stability [61, 74, 135].
- $Q(z)$ is designed as a low-pass zero-phase FIR filter of order 200 with cut-off frequency 620 Hz.

Noncausality of filters $L_h(z)$, $L_v(z)$ and $Q(z)$ is accounted for as indicated in Section 5.2.2. Including the inverse plant model in $L(z)$ guarantees nominal stability, since in case of perfect plant models $\widehat{G}_h(z)$ and $\widehat{G}_v(z)$ and $F(z) = Q(z) = 1$, $S_h(z)$ and $S_v(z)$ are both equal to:

$$\overline{S}(z) = 1 - \chi(z) . \qquad (7.7)$$

However, the identified decoupled plant model $\widehat{\mathbf{G}}(z)$, Equation 7.1, is only accurate up to 600 Hz, raising the issue of robust stability. To this end, low-pass filters $Q(z)$ and $F(z)$ turn off the repetitive controller from 600 Hz. Their orders and cut-off frequencies are tuned such that the following sufficient MIMO stability criterion (derived from the small gain theorem, similar to [61]) is satisfied for the three designs of $\chi(z)$ selected in Section 7.3.2:

$$\sup_{\omega \in \mathbf{R}} \left\{ \max_i \{|\lambda_i\left(H(\omega)\right)|\} \right\} < \frac{1}{\|\chi(z)\|_\infty}, \tag{7.8}$$

where

$$H(z) = \begin{bmatrix} \left(1 - G_h(z)\widehat{G}_h(z)^{-1}F(z)\right)Q(z) & -G_{hv}(z)\widehat{G}_v(z)^{-1}F(z)Q(z) \\ -G_{vh}(z)\widehat{G}_h(z)^{-1}F(z)Q(z) & \left(1 - G_v(z)\widehat{G}_v(z)^{-1}F(z)\right)Q(z) \end{bmatrix},$$

and $\lambda_i(X)$ denote the eigenvalues of X. The left-hand side of (7.8) is computed from the nonparametric FRF estimates of $\mathbf{G}(z)$.

Performance

According to Section 5.3.1, performance is specified as an optimal trade-off between two performance indices that quantify the closed-loop steady-state performance with respect to periodic and nonperiodic inputs. The definitions of these performance indices are simplified by the assumptions that in the pass band of the filters $Q(z)$ and $F(z)$ (i) the filters equal their dc-gain; and (ii) the identified plant models are perfect. These assumptions imply that $S_h(\omega) = S_v(\omega) = \overline{S}(\omega)$ holds up to 600 Hz.

According to (5.11b), periodic performance index γ_p corresponds to the smallest reduction $|\overline{S}(l\omega_{p,\delta})|$ over all harmonics $l \in \mathcal{L}$ and over all potential $\omega_{p,\delta}$ values. For the control problem at hand: $\mathcal{L} = \{1, 2, \ldots, 30\}$, where $l_{max} = 30$ follows from the fact that the repetitive controller action is restricted to 600 Hz, which corresponds to the 30'th harmonic at 1200 rpm, the highest spindle rotational speed. As all harmonics except $l = 1$ have a similar contribution to $\mathbf{e_d}(k)$ (see Figure 7.10(a)), weights $W_l = 1$ are used $\forall l \in \mathcal{L}$. Hereby, definition (5.11b) of γ_p reduces to

$$\gamma_p = \max_{\omega \in \overline{\Omega}_{l_{max}}} \{|\overline{S}(\omega)|\},$$

where $\overline{\Omega}_l$ is given by (5.12).

By adopting (5.10b), nonperiodic performance index γ_{np} corresponds to the highest amplification $|\overline{S}(\omega)|$ over all frequencies ω:

$$\gamma_{np} = \|\overline{S}(z)\|_\infty.$$

The repetitive controller is designed to yield an optimal trade-off between the conflicting performance indices γ_p and γ_{np}. To that end, parameters χ_m are computed as the solution of the following optimization problem, for given $\alpha \geq 0$:

$$\underset{\chi_m, \gamma_p, \gamma_{np}}{\text{minimize}} \quad \gamma_p + \alpha\gamma_{np} \tag{7.9a}$$

$$\text{subject to} \quad \|\bar{S}(z)\|_\infty \leq \gamma_{np} \tag{7.9b}$$

$$|\bar{S}(\omega)| \leq \gamma_p, \quad \forall\omega \in \overline{\Omega}_{l_{max}}. \tag{7.9c}$$

7.3.2 Selected Repetitive Controllers

As explained in Section 5.3.1, the solution of (7.9) does not depend on f_p, while it only depends on $\mathscr{L}$ and δ through the product $l_{max}\delta$. In the repetitive controller design $l_{max}\delta$ is set equal to 2%, corresponding to the initial estimate $\delta = 0.07\%$.

Three repetitive controllers, characterized by a different design of $\chi(z)$, are experimentally validated. The first repetitive controller, denoted by $\mathbf{K_{RC1}}(z)$, corresponds to the more frequently used basic first-order repetitive controller $\chi(z) = z^{-N}$. This controller yields $\gamma_{np} = 2$ and, although this controller perfectly rejects periodic disturbances at f_p, its robust periodic performance is moderate: $\gamma_p = 0.13$.

The Pareto optimal repetitive controllers $\mathbf{K_{RC2}}(z)$ and $\mathbf{K_{RC3}}(z)$, on the other hand, are fifth-order repetitive controllers ($\mu = 5$), optimized according to (7.9) and corresponding to different trade-offs between γ_p and γ_{np}, that is: different weights α in (7.9). Figure 7.7 indicates both designs on the trade-off curve between γ_p and γ_{np} for fifth-order repetitive controllers, which is computed by solving (7.9) with $\mu = 5$ for a range of α values. For a given level of periodic performance γ_p, the trade-off curve indicates the minimal level of nonperiodic performance degradation γ_{np} that has to be tolerated. Or, *vice versa*, for a fixed level of nonperiodic performance, the trade-off curve indicates the best periodic performance that can be achieved. The steep slope between $\mathbf{K_{RC2}}(z)$ and $\mathbf{K_{RC3}}(z)$ indicates that improving the periodic performance below $\gamma_p = 0.022$ comes at the price of high amplification of nonperiodic disturbances: compared to $\mathbf{K_{RC2}}(z)$, $\mathbf{K_{RC3}}(z)$ improves the periodic performance from $\gamma_p = 0.022$ to $\gamma_p = 0.0013$, but degrades the nonperiodic performance from $\gamma_{np} = 1.8$ to $\gamma_{np} = 3.3$. The first-order repetitive controller $\mathbf{K_{RC1}}(z)$ is also indicated in Figure 7.7, but as it does not correspond to an optimal fifth-order design, it is located off the trade-off curve.

Figure 7.8 compares, for the considered repetitive controller designs, the FRF of $\bar{S}(z)$, which is a good approximation of both $S_h(\omega)$ and $S_v(\omega)$ up to 600 Hz. Since $\chi(z)$, Equation 7.5, contains only powers of z^{-N}, the FRF of $\bar{S}(z)$, Equation 7.7, is periodic with ω_p. Therefore, Figure 7.8 only shows this FRF for the frequency range $[0, \omega_p]$. γ_{np} is found as the peak value of $|\bar{S}(\omega)|$ over the entire frequency range, whereas γ_p corresponds to the maximum of $|\bar{S}(\omega)|$ over the 2% uncertainty interval, indicated by the shaded band.

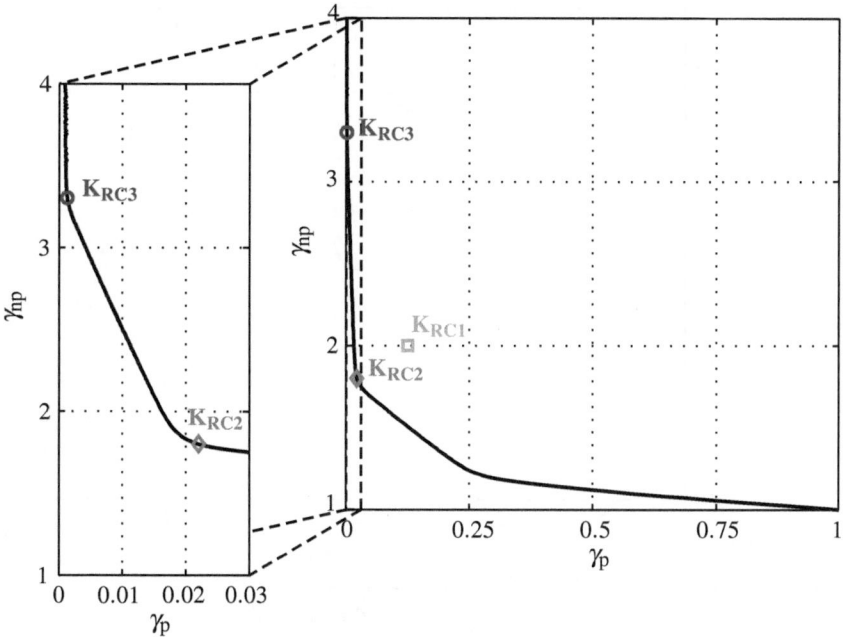

Fig. 7.7 Trade-off curve between γ_p and γ_{np} for fifth-order repetitive controllers, where the selected controller designs are indicated: $\mathbf{K_{RC2}}(z)$ and $\mathbf{K_{RC3}}(z)$ correspond to Pareto optimal controllers, whereas $\mathbf{K_{RC1}}(z)$ is a classical first-order repetitive controller.

7.4 Experimental Results

The three selected repetitive controllers are experimentally validated on the active air bearing setup. First, the controllers are validated for a fixed rotational speed of 1200 rpm, yielding $f_p = f_{p,des} = 20$ Hz and $N = 500$. Sections 7.4.1 and 7.4.2 respectively assess the corresponding transient and steady-state performance of the controllers.

Second, the adaptive implementation of the repetitive controllers is experimentally validated (Section 7.4.3), which allows dealing with large variations of the spindle's rotational speed setpoint, as occurring during run-ups and run-downs. Following [42, 139], N, the only parameter that depends on the spindle's rotational speed, is adapted according to its estimate obtained from the index pulse (one pulse per revolution) of the encoder.

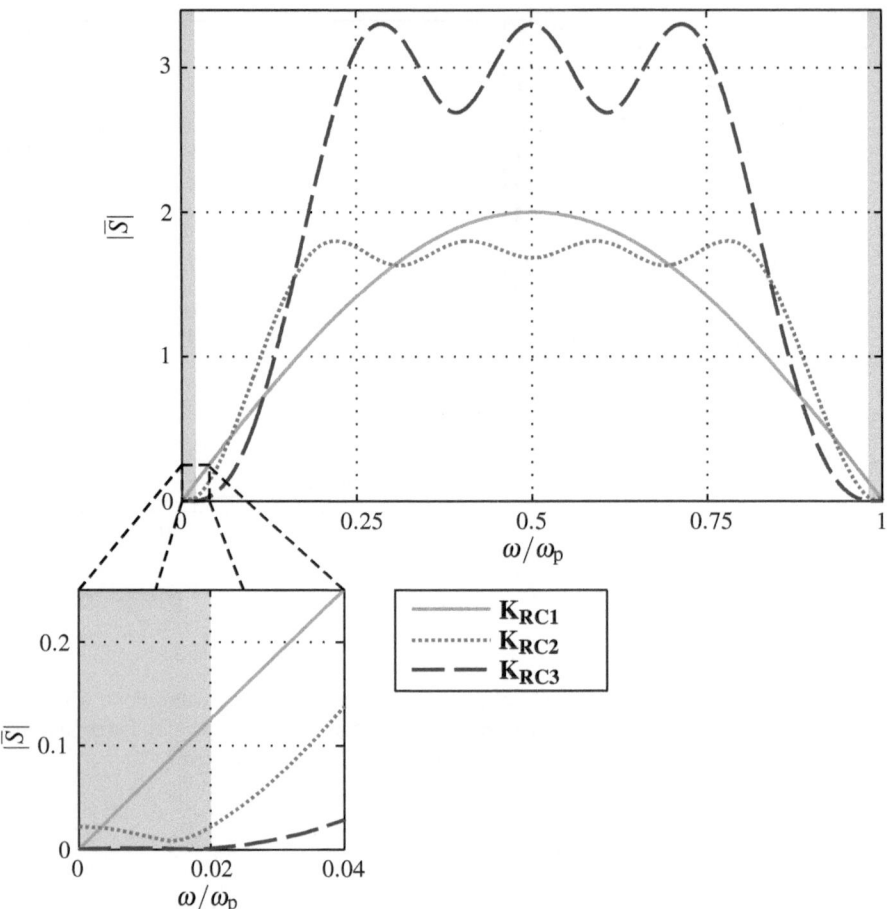

Fig. 7.8 Comparison of $\overline{S}(z)$, defined as the closed-loop sensitivity in the case of a perfect plant model and $F(z) = Q(z) = 1$, for the selected repetitive controller designs. Since these FRFs are periodic with ω_p, they are only shown for the frequency range $[0, \omega_p]$. The shaded bands indicate the $l_{max}\boldsymbol{\delta} = 2\%$ uncertainty interval.

7.4.1 Transient Response (1200 rpm)

To evaluate the transient response of the controllers, Figure 7.9 shows the measured horizontal error motion $e_{m,h}(k)$ of the spindle's axis of rotation, where the repetitive controllers are switched on at $t = 0.5$ s. In contrast to the following sections, which show results for the decoupled controllers (7.2), Figure 7.9 relates to the situation where only the horizontal SISO controllers $K_{RC1,h}(z)$, $K_{RC2,h}(z)$ and $K_{RC3,h}(z)$ are switched on, while the vertical control loop is left open.

Whereas the first-order repetitive controller $K_{RC1,h}(z)$ needs only one period for reaching steady state, the transient response of the fifth-order repetitive controllers

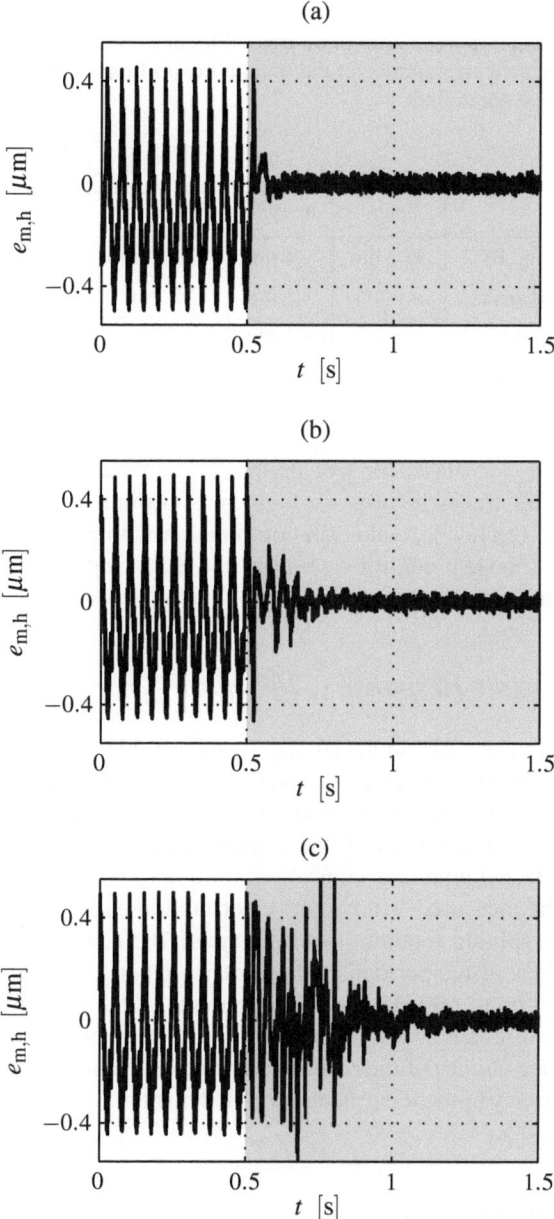

Fig. 7.9 Transient response (1200 rpm): measured horizontal error motion $e_{m,h}(k)$ of the spindle's axis of rotation, where the repetitive controllers are switched on at $t = 0.5$ s; results for (a) $K_{RC1,h}(z)$; (b) $K_{RC2,h}(z)$; and (c) $K_{RC3,h}(z)$.

Table 7.1 Steady-state response (1200 rpm): rms values of the periodic and nonperiodic contribution to the measured error motion of the spindle's axis of rotation in the active frequency range of the repetitive controllers.

	horizontal		vertical	
	per.	nonper.	per.	nonper.
no RC	95.2 nm	2.4 nm	113.4 nm	2.9 nm
$\mathbf{K_{RC1}}(z)$	4.3 nm	2.1 nm	3.9 nm	3.9 nm
$\mathbf{K_{RC2}}(z)$	2.7 nm	2.1 nm	2.7 nm	4.2 nm
$\mathbf{K_{RC3}}(z)$	0.7 nm	4.4 nm	0.9 nm	8.6 nm

$K_{RC2,h}(z)$ and $K_{RC3,h}(z)$ lasts for five periods. Compared to $K_{RC2,h}(z)$, $K_{RC3,h}(z)$ yields a more gruff transient response, which is attributed to its worse nonperiodic performance (higher γ_{np} value) in combination with actuator saturation: only for $K_{RC3,h}(z)$ the transient control signal hits the ± 2 V input bounds of the piezo-actuator amplifier.

7.4.2 Steady-state Response (1200 rpm)

Due to the Bode Integral Theorem [10, 21, 22, 49, 69, 138], a repetitive controller's steady-state performance is bound to a trade-off between the suppression of periodic disturbances and the amplification of nonperiodic inputs. This section experimentally investigates the implications of this performance trade-off on the test setup by comparing the selected repetitive controllers.

To this end, the measured steady-state error motion $\mathbf{e_m}(k)$ is split up into a part periodic with the spindle rotation, and its nonperiodic content. Table 7.1 summarizes the rms values of the periodic and nonperiodic part of $\mathbf{e_m}(k)$, where only the frequency content up to 600 Hz, the working range of the repetitive controllers, is accounted for. As discussed below, Table 7.1 can to a large extent be explained based on the performance indices of the repetitive controller. Hence, the experiments support both the accuracy of the identified decoupled plant model $\widehat{\mathbf{G}}(z)$ and the practical relevance of γ_p and γ_{np}.

Periodic Performance

To allow for a detailed evaluation of the repetitive controllers' periodic performance, Figure 7.10 shows the amplitude spectrum of the periodic part of the measured horizontal error motion $e_{m,h}(k)$ up to the 30'th harmonic, and the corresponding reduction achieved by the repetitive controllers. The actual period of the spindle rotation is estimated *a posteriori* using the approach of [124], yielding $f_{p,\delta} \cong 19.968$ Hz and hence, $\delta = 0.16\%$.

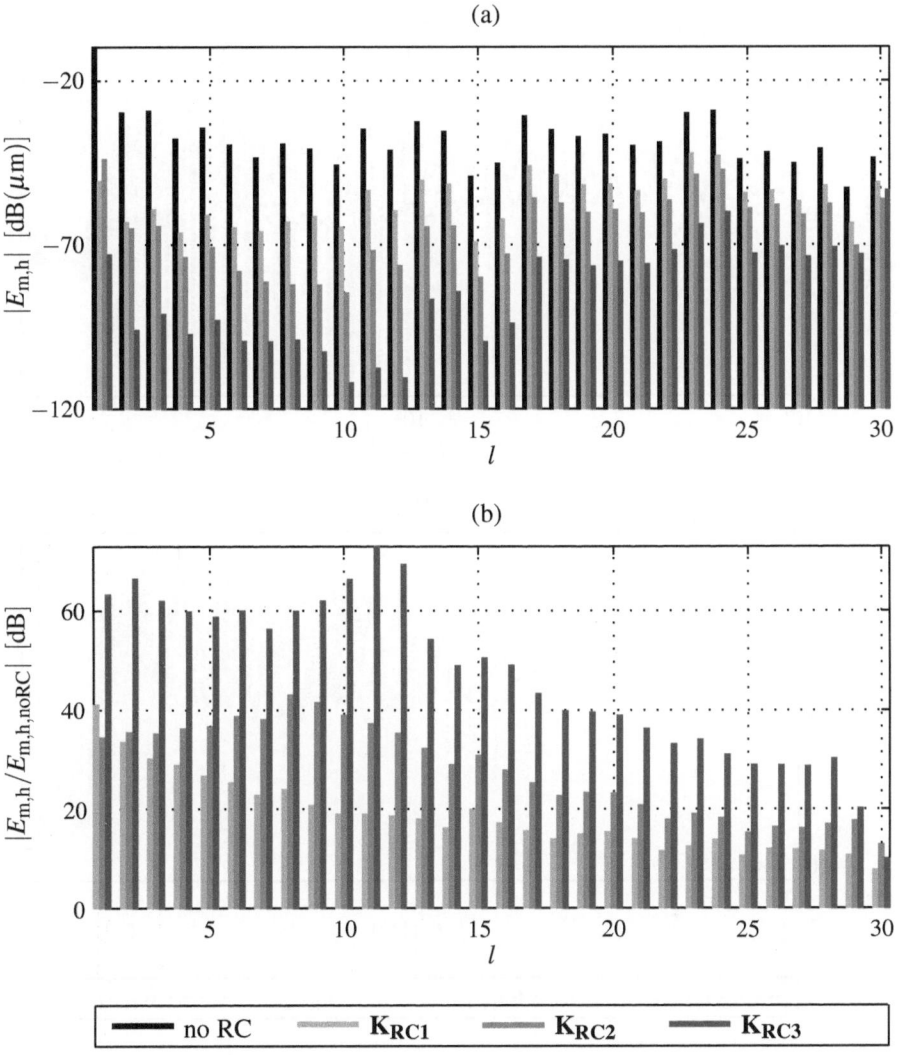

Fig. 7.10 Periodic steady-state response (1200 rpm): (a) amplitude spectrum of the periodic part of the measured horizontal error motion $e_{m,h}(k)$; and (b) the corresponding reduction achieved by the three repetitive controllers.

Converted to dB, the γ_p values imply that all harmonics up to $l = \mathrm{int}(\boldsymbol{\delta}/\delta) = \mathrm{int}(2/0.16) = 12$, are reduced 18 dB by $\mathbf{K}_{RC1}(z)$, 33 dB by $\mathbf{K}_{RC2}(z)$, and 58 dB by $\mathbf{K}_{RC3}(z)$; which is confirmed by Figure 7.10(b). Furthermore, both Figure 7.10 and Table 7.1 indicate that a lower γ_p value gives rise to a smaller periodic contribution to $\mathbf{e}_m(k)$, even if 30 harmonics are taken into account.

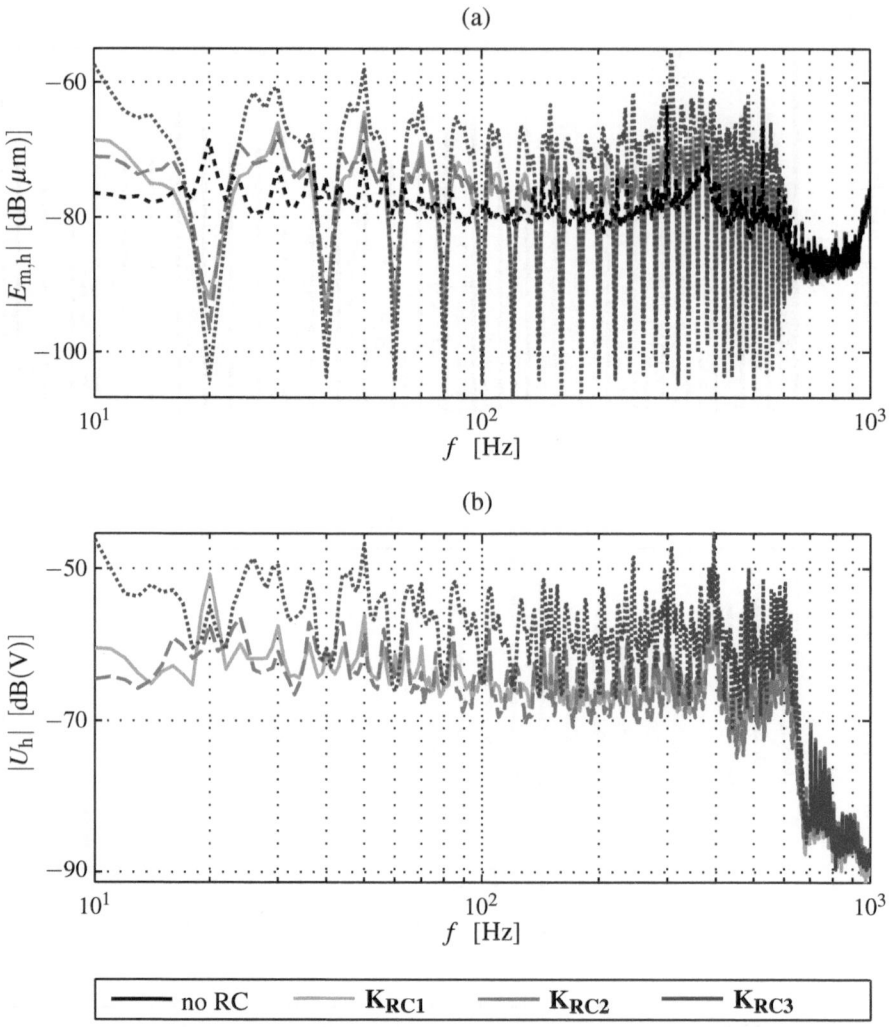

Fig. 7.11 Nonperiodic steady-state response (1200 rpm): amplitude spectrum of the nonperiodic part of (a) the measured horizontal error motion $e_{m,h}(k)$; and (b) the horizontal control signal $u_h(k)$ for the three repetitive controllers.

The results for $\mathbf{K}_{RC1}(z)$ reveal the necessity to account for uncertainty on f_p. Whereas this controller would perfectly eliminate the periodic contribution to $\mathbf{e}_m(k)$ if $\delta = 0\%$, the small deviation $\delta = 0.16\%$ causes a significant periodic performance degradation for $\mathbf{K}_{RC1}(z)$.

Nonperiodic Performance

For the control problem at hand, the measurement noise $\mathbf{n}(k)$ constitutes the sole nonperiodic input to the closed-loop system. Figure 7.11 complements the results of Table 7.1 by showing the amplitude spectrum of the nonperiodic part of the measured error motion $\mathbf{e_m}(k)$ and the control signal $\mathbf{u}(k)$.

Interpretation of the results of Figure 7.11(a) and Table 7.1 requires special care, for they involve the measured error motion $\mathbf{e_m}(k)$. While $\mathbf{n}(k)$ relates to $\mathbf{e_m}(k)$ by the closed-loop sensitivity, its effect on the actual error motion $\mathbf{e}(k)$ is determined by the complementary sensitivity. For this reason, Figure 7.11(a) and Table 7.1 only provide an indication of the measurement noise contribution to the error motion $\mathbf{e}(k)$. However, main conclusions regarding the nonperiodic performance remain valid: $\mathbf{K_{RC1}}(z)$ and $\mathbf{K_{RC2}}(z)$ yield a similar, modest, nonperiodic contribution to $\mathbf{e}(k)$, whereas $\mathbf{K_{RC3}}(z)$ amplifies the measurement noise with a factor of almost two. This is also clear from the effect of the measurement noise on the control signal, see Figure 7.11(b), since this nonperiodic contribution is transmitted to the system, thereby directly affecting the output $\mathbf{e_u}(k)$ and the actual error motion $\mathbf{e}(k)$. The simulated response of $\widehat{\mathbf{G}}(z)$ to the experimental control signals yields similar results as Figure 7.11(a) and Table 7.1.

Inspection of the rms values in Table 7.1 reveals that the superior periodic performance of $\mathbf{K_{RC3}}(z)$ comes at too high an amplification of the measurement noise, deteriorating the overall closed-loop performance. From this point of view $\mathbf{K_{RC2}}(z)$ is preferred, as it combines better periodic performance than $\mathbf{K_{RC1}}(z)$ with a similar nonperiodic contribution to $\mathbf{e}(k)$.

7.4.3 Adaptive Response (900–1200 rpm)

The adaptive implementation of the repetitive controllers is validated for a run-down from 1200 rpm to 900 rpm in 9.85 s and a run-up from 900 rpm to 1200 rpm in the same time frame. Figure 7.12(a) shows the relative rotational speed variation corresponding to the constant deceleration profile entered to the frequency converter:

$$\alpha_{p,des} = \frac{1}{f_{p,des}} \frac{df_{p,des}}{dt} .$$

The horizontal error motion $e_{m,h}(k)$ of the axis of rotation measured during this run-down without control is shown in Figure 7.12(b), while Figure 7.13 shows the results for the three selected repetitive controllers. To quantify the adaptive performance of the controllers, Table 7.2 summarizes the rms values of the measured error motion $\mathbf{e_m}(k)$ during the 9.85 s run-down and run-up.

To which extent the repetitive controllers preserve their steady-state performance depends on two factors. The first and dominating factor is $\alpha_{p,des}$: while small relative speed variations are accounted for by the robustness of the repetitive

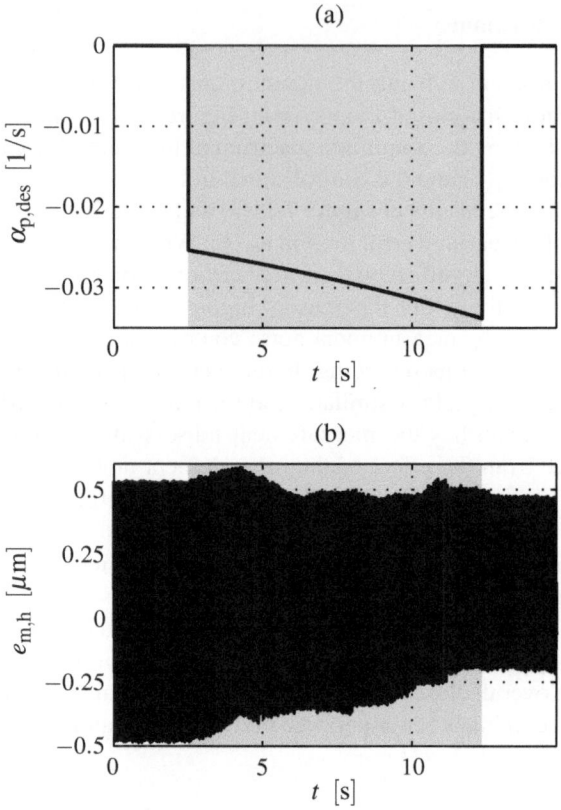

Fig. 7.12 Adaptive response (900–1200 rpm): (a) relative speed variation $\alpha_{p,des}$; and (b) measured horizontal error motion of the axis of rotation without control, for a run-down from 1200 rpm to 900 rpm in 9.85 s.

Table 7.2 Adaptive response (900–1200 rpm): rms values of the measured error motion of the spindle's axis of rotation during a run-down and run-up between 1200 rpm and 900 rpm, each in 9.85 s.

	horizontal		vertical	
	run-down	run-up	run-down	run-up
no RC	80.0 nm	81.0 nm	95.7 nm	96.0 nm
$\mathbf{K_{RC1}}(z)$	9.0 nm	8.6 nm	9.4 nm	9.5 nm
$\mathbf{K_{RC2}}(z)$	7.9 nm	7.6 nm	8.6 nm	8.9 nm
$\mathbf{K_{RC3}}(z)$	10.2 nm	10.7 nm	14.1 nm	13.7 nm

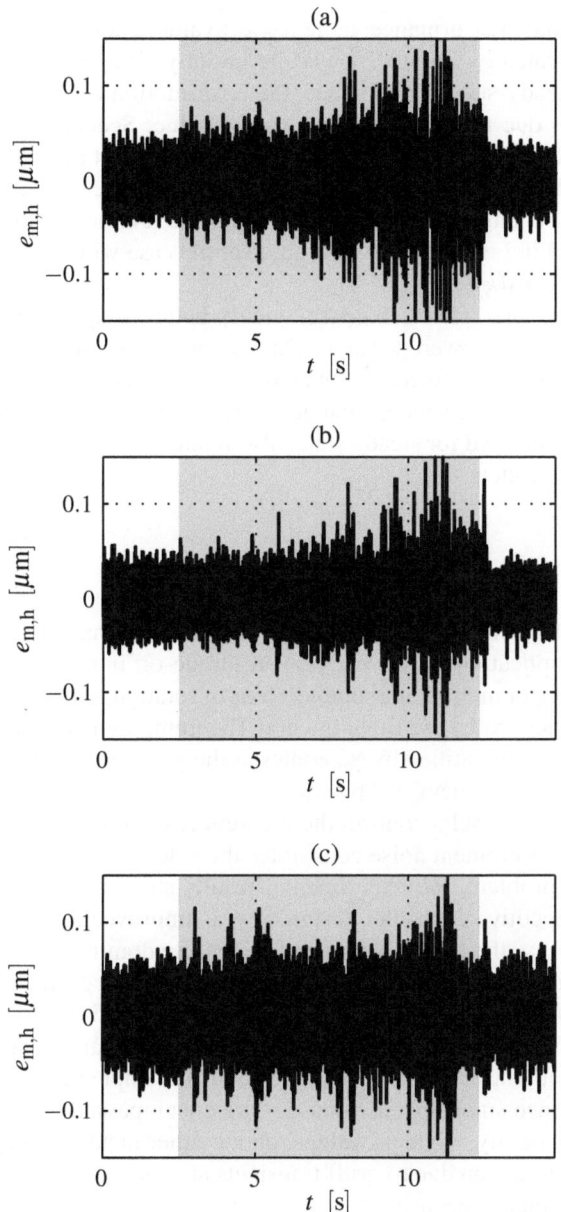

Fig. 7.13 Adaptive response (900–1200 rpm): measured horizontal error motion of the axis of rotation for a run-down from 1200 rpm to 900 rpm in 9.85 s; results for (a) $\mathbf{K_{RC1}}(z)$; (b) $\mathbf{K_{RC2}}(z)$; and (c) $\mathbf{K_{RC3}}(z)$.

controller's periodic performance, large $|\alpha_{p,des}|$ values cause performance degradation. This is revealed by Figure 7.13: while initially all three repetitive controllers preserve their steady-state performance, the error motion deteriorates near the end of the run-down due to the increasing $|\alpha_{p,des}|$ value. Second, sudden changes in the amplitude/phase spectrum of the error motion $\mathbf{e_d}(k)$ put the higher-order repetitive controllers $\mathbf{K_{RC2}}(z)$ and $\mathbf{K_{RC3}}(z)$ at a disadvantage due to their longer transient response to these changes. These sudden changes are revealed by the nonsmooth envelope of $e_{m,h}(k)$ in Figure 7.12(b), because for the case without control: $\mathbf{e_u}(k) = \mathbf{0}$ and hence, $\mathbf{e}(k) = \mathbf{e_d}(k)$.

Related to the first factor, $\mathbf{K_{RC2}}(z)$ yields better adaptive performance than $\mathbf{K_{RC1}}(z)$ thanks to its lower γ_p value. On account of the second factor, $\mathbf{K_{RC3}}(z)$ yields worse adaptive performance than $\mathbf{K_{RC2}}(z)$, as its high γ_{np} value translates into gruff transients at the sudden changes in $\mathbf{e_d}(k)$. Hence, the performance indices γ_{np} and γ_p, being defined for steady state, also relate to the adaptive performance of the repetitive controllers.

7.5   Conclusion

Relying on the repetitive controller design proposed in Chapter 5, this chapter investigates the implications of the performance trade-off in repetitive control for the reduction of the error motion of a spindle's axis of rotation, supported in an active air bearing setup. Dictated by the Bode Integral Theorem, improved suppression of periodic disturbances, quantified by γ_p, comes at the price of a degraded performance for nonperiodic inputs, quantified by γ_{np}.

The experimental results confirm the theoretical trade-off curve between γ_p and γ_{np}. Although measurement noise constitutes the sole nonperiodic input to the considered control problem, the experimental results show that it should not be neglected in the repetitive controller design. If γ_p is improved at the price of too high a degradation of γ_{np}, the fed-back measurement noise dominates the error motion of the axis of rotation. Moreover, high γ_{np} values give rise to a gruff transient response.

Although defined for steady state, the performance indices γ_p and γ_{np} also relate to the repetitive controller's adaptive performance during large variations of the spindle's rotational speed. The adaptive controller implementation benefits from low γ_p values, since small speed variations are encompassed by the robustness for period-time uncertainty. High γ_{np} values, on the other hand, translate into adaptive performance degradation due to gruff transients at sudden changes in the error motion's amplitude/phase spectrum.

Chapter 8
Conclusions

This monograph presents a general design methodology for linear controllers facing periodic inputs. As every rotating machine and repeated process involves periodicity, periodic signals are widespread in engineering practice. Consequently, control for periodic inputs has gained a marked status in modern control literature, where four control strategies are distinguished: feedforward control, estimated disturbance feedback control, repetitive control and feedback control. The presented design methodology applies to these four controller types, and in all cases it is able to reproduce and outperform major current design strategies. The superior performance of the proposed methodology stems from the following properties:

Multi-objective Control: The majority of existing design approaches enforce perfect closed-loop periodic performance, that is: perfect asymptotic tracking/rejection of the periodic input, without investigating the corresponding degradation of alternative closed-loop performance aspects. The proposed design methodology, on the other hand, adopts a multi-objective design philosophy, and although improving closed-loop periodic performance remains the primal objective, it is traded-off against other performance specifications.

Period-time Uncertainty: The majority of existing design approaches cannot cope with period-time uncertainty and hence, inherently assume the input period to be accurately known or measurable. The developed design methodology, on the other hand, provides a systematic way to handle this uncertainty: instead of accounting for the nominal period-time only, the methodology introduces the periodic performance index, which quantifies the worst-case closed-loop periodic performance over all potential values of the input period. The advantage of the periodic performance index is illustrated by numerical results and its practical relevance is sustained by experiments.

Convex Optimization: To guarantee a reliable and efficient computation of the global optimum, the multi-objective controller design problem is translated into a convex optimization problem. This transformation is enabled by the Youla parametrization, while further manipulations are generally required to render the optimization problem numerically tractable. This latter step can be accomplished

G. Pipeleers et al.: Optimal Linear Controller Design for Periodic Inputs, LNCIS 394, pp. 141–142.
springerlink.com   © Springer-Verlag Berlin Heidelberg 2009

by either gridding or application of the (generalized) KYP lemma, where the advantages of both approaches are illustrated.

Thanks to these properties, the design methodology is innovative with respect to the current literature. In addition, the design methodology achieves the following contributions:

Mutual Relations: The presented methodology can be translated into a feedforward, estimated disturbance feedback, repetitive or feedback controller design and hereby emphasizes their mutual relations. The major differences between feedforward and estimated disturbance feedback control are revealed, and when applying the methodology to the feedback controller design, the relations with repetitive and estimated disturbance feedback control are highlighted.

Limits of Performance: In feedback control, the Bode Integral Theorem dictates a fundamental trade-off between periodic performance improvement, nonperiodic performance degradation and duration of the transient response. The proposed design methodology allows a systematic and quantitative analysis of this trade-off by translating it into a trade-off surface between the periodic performance index, the nonperiodic performance index and the finite impulse response length of the (modifying) sensitivity. The computation of this trade-off surface is facilitated by the convex reformulation of the controller design problem.

Performance of Repetitive Controllers: Repetitive controllers are feedback controllers that only exploit the input periodicity, while more general feedback controllers additionally account for the input's harmonic frequency content. Although repetitive controllers benefit from a simple structure and intuitive design, their particular structure impedes repetitive controllers to optimize closed-loop periodic performance properly. In the general feedback controller design, the Youla parametrization is chosen such that repetitive controllers are encompassed as a special case, and hereby the performance loss caused by the repetitive controller structure is analyzed in view of the aforementioned performance trade-off.

Appendix A
Semi-definite Programming Reformulation of Optimal Controller Design

A.1 Introduction

This appendix details how to transform the semi-infinite constraints that define γ_p, Equation 2.7, and γ_{np}, Equation 2.10, into LMIs, and derives three equivalent formulations for the resulting SDP. These SDP reformulations are presented for the particular optimization problem discussed in Section 2.3.3 for SISO systems $H_{np}(z)$ and $H_p(z)$, while extending these results to MIMO systems and alternative design problems is rather straightforward. Hence, this appendix deals with the following semi-infinite optimization problem:

$$\underset{x,\gamma_p,\gamma_{np}}{\text{minimize}} \quad \gamma_p + \alpha \gamma_{np} \tag{A.1a}$$

$$\text{subject to} \quad \|H_{np}(z,x)\|_\infty \le \gamma_{np} \tag{A.1b}$$

$$W_l |H_p(\omega,x)| \le \gamma_p, \quad \forall \omega \in \Omega_l, \quad \forall l \in \mathscr{L}, \tag{A.1c}$$

where optimization variable $x \in \mathbf{R}_M$ is added as an argument in $H_{np}(z,x)$ and $H_p(z,x)$ to indicate relations (2.13) and (2.15) corresponding to the Youla parametrization. These relations allow for state-space models in the following form:

$$H_p(z,x) = C_p(x)(zI - A)^{-1}B_p + D_p(x),$$
$$H_{np}(z,x) = C_{np}(x)(zI - A)^{-1}B_{np} + D_{np}(x),$$

where the relations $C_p(x)$, $D_p(x)$, $C_{np}(x)$ and $D_{np}(x)$ are affine in x:

$$\left[C_p(x)\ D_p(x)\right]^T = E_p x + F_p,$$
$$\left[C_{np}(x)\ D_{np}(x)\right]^T = E_{np} x + F_{np}.$$

The order of closed-loop systems $H_p(z,x)$ and $H_{np}(z,x)$ is denoted by n and depends affinely on M.

The reformulation of the optimal design problem into an SDP relies on the KYP lemma [79, 113, 157] and the generalized KYP lemma [77, 123]. Section A.2 briefly reviews these lemmas and applies them to the constraints of (A.1). Solving the resulting SDP with a standard interior-point solver involves a computational complexity of $\mathcal{O}(n^6)$. Section A.3 elaborates the equivalent reformulation of this SDP proposed by Vandenberghe *et al.* [150], which is solved with a standard interior-point solver at the cost of $\mathcal{O}(n^4)$. Subsequently, Section A.4 presents the SDP reformulation derived by Liu and Vandenberghe [94], which can be solved with computational complexity $\mathcal{O}(n^3)$ by exploiting its structure.

The computational complexities indicated in this appendix in fact correspond to the cost of one iteration of an interior-point solver. Since in general, the number of iterations of an interior-point solver grows very slowly with the problem size [15], these computational complexities provide an accurate estimate of the overall cost of solving the SDP. In addition, the affine dependency $M = \mathcal{O}(n)$ is exploited in the complexity analysis, and $n_{\mathscr{L}} \ll n$ is assumed.

A.2   Application of the (Generalized) KYP Lemma

A.2.1   KYP Lemma and Its Generalization

For a discrete-time system $H(z)$, the KYP lemma and its generalization analyze the following inequality:

$$\begin{bmatrix} H(z) \\ I \end{bmatrix}^H \Pi \begin{bmatrix} H(z) \\ I \end{bmatrix} \le 0 , \tag{A.3}$$

on the unit circle, where Π is a given symmetric matrix. The order of $H(z)$ is denoted by n, and matrices A, B, C and D correspond to a state-space model of $H(z)$:

$$H(z) = C(zI - A)^{-1}B + D .$$

Although not essential to the (generalized) KYP lemma, matrix A is assumed to be stable and the pair (A, B) controllable.

The KYP lemma [79, 113, 157] considers inequality (A.3) over the entire frequency range:

$$z \in \left\{ e^{j\omega T_s} \,\middle|\, \omega \in \mathbf{R} \right\} ,$$

and this subset of $\mathbf{C}$ corresponds to

$$\varUpsilon(\varPhi, 0) = \left\{ z \in \mathbf{C} \,\middle|\, \begin{bmatrix} z \\ 1 \end{bmatrix}^H \underbrace{\begin{bmatrix} 1 & 0 \\ 0 & -1 \end{bmatrix}}_{\varPhi} \begin{bmatrix} z \\ 1 \end{bmatrix} = 0 \right\} .$$

The KYP lemma states the equivalence between the following statements:

- Frequency domain inequality (A.3) holds for all $z \in \Upsilon(\Phi, 0)$.
- There exists a matrix $P \in \mathbf{S}_n$ that satisfies the following LMI:

$$\begin{bmatrix} A & B \\ I & 0 \end{bmatrix}^T (\Phi \otimes P) \begin{bmatrix} A & B \\ I & 0 \end{bmatrix} + \begin{bmatrix} C & D \\ 0 & I \end{bmatrix}^T \Pi \begin{bmatrix} C & D \\ 0 & I \end{bmatrix} \preceq 0, \tag{A.4}$$

where $\otimes$ indicates the matrix Kronecker product.

The generalized KYP lemma [77, 123] considers inequality (A.3) over a limited frequency range:

$$z \in \left\{ e^{j\omega T_s} \mid \omega \in [\omega_1, \omega_2] \right\}.$$

This subset of $\mathbf{C}$ corresponds to

$$\Upsilon(\Phi, \Psi) = \left\{ z \in \mathbf{C} \,\middle|\, \begin{bmatrix} z \\ 1 \end{bmatrix}^H \Phi \begin{bmatrix} z \\ 1 \end{bmatrix} = 0; \; \begin{bmatrix} z \\ 1 \end{bmatrix}^H \Psi \begin{bmatrix} z \\ 1 \end{bmatrix} \geq 0 \right\},$$

where

$$\Psi = \begin{bmatrix} 0 & \exp(j\omega_c T_s) \\ \exp(-j\omega_c T_s) & -2\cos(\omega_d T_s) \end{bmatrix},$$

with

$$\omega_c = 0.5(\omega_1 + \omega_2),$$
$$\omega_d = 0.5(\omega_2 - \omega_1).$$

The generalized KYP lemma states the equivalence between the following statements:

- Frequency domain inequality (A.3) holds for all $z \in \Upsilon(\Phi, \Psi)$.
- There exist matrices $P, Q \in \mathbf{H}_n$ that satisfy the following set of LMIs:

$$Q \succeq 0, \tag{A.5a}$$

$$\begin{bmatrix} A & B \\ I & 0 \end{bmatrix}^T (\Phi \otimes P + \Psi \otimes Q) \begin{bmatrix} A & B \\ I & 0 \end{bmatrix} + \begin{bmatrix} C & D \\ 0 & I \end{bmatrix}^T \Pi \begin{bmatrix} C & D \\ 0 & I \end{bmatrix} \preceq 0. \tag{A.5b}$$

A.2.2 Primal SDP

Applying the KYP lemma to constraint (A.1b) yields the following LMI in $x \in \mathbf{R}_M$, $\gamma_{np} \in \mathbf{R}$ and $P_{np} \in \mathbf{S}_n$:

$$\mathcal{K}_{np}(P_{np}) + \mathcal{E}_{np}(x) + \mathcal{G}_{np}(\gamma_{np}) + \mathcal{F}_{np} \preceq 0,$$

where matrix $\mathcal{F}_{np}$ and linear mappings $\mathcal{K}_{np}(P_{np})$, $\mathcal{E}_{np}(x)$ and $\mathcal{G}_{np}(\gamma_{np})$ are defined as follows:

$$\mathscr{F}_{np} = \left[\begin{array}{cc|c} 0 & 0 & F_{np} \\ 0 & 0 & \\ \hline F_{np}^T & & 0 \end{array}\right],$$

$$\mathscr{K}_{np}(P_{np}) = \left[\begin{array}{ccc} A^T P_{np} A - P_{np} & A^T P_{np} B_{np} & 0 \\ B_{np}^T P_{np} A & B_{np}^T P_{np} B_{np} & 0 \\ 0 & 0 & 0 \end{array}\right],$$

$$\mathscr{E}_{np}(x) = \left[\begin{array}{cc|c} 0 & 0 & E_{np} x \\ 0 & 0 & \\ \hline (E_{np} x)^T & & 0 \end{array}\right],$$

$$\mathscr{G}_{np}(\gamma_{np}) = \left[\begin{array}{ccc} 0 & 0 & 0 \\ 0 & -\gamma_{np} & 0 \\ 0 & 0 & -\gamma_{np} \end{array}\right].$$

The generalized KYP lemma replaces each of the constraints (A.1c) by the following set of LMIs in $x \in \mathbf{R}_M$, $\gamma_p \in \mathbf{R}$ and $P_{p,l}, Q_{p,l} \in \mathbf{H}_n$:

$$Q_{p,l} \succeq 0,$$

$$\mathscr{K}_p(P_{p,l}) + \mathscr{R}_{p,l}(Q_{p,l}) + \mathscr{E}_p(x) + \mathscr{G}_p(\gamma_p) + \mathscr{F}_p \preceq 0,$$

where matrix $\mathscr{F}_p$ and linear mappings $\mathscr{K}_p(P_{p,l})$, $\mathscr{R}_{p,l}(Q_{p,l})$, $\mathscr{E}_p(x)$ and $\mathscr{G}_p(\gamma_p)$ are defined as follows:

$$\mathscr{F}_p = \left[\begin{array}{cc|c} 0 & 0 & F_p \\ 0 & 0 & \\ \hline F_p^T & & 0 \end{array}\right],$$

$$\mathscr{K}_p(P_{p,l}) = \left[\begin{array}{ccc} A_p^T P_{p,l} A_p - P_{p,l} & A_p^T P_{p,l} B_p & 0 \\ B_p^T P_{p,l} A_p & B_p^T P_{p,l} B_p & 0 \\ 0 & 0 & 0 \end{array}\right],$$

$$\mathscr{R}_{p,l}(Q_{p,l}) = \left[\begin{array}{ccc} \eta_l A_p^T Q_{p,l} + \eta_l^H Q_{p,l} A_p + \zeta_l Q_{p,l} & \eta_l^H Q_{p,l} B_p & 0 \\ \eta_l B_p^T Q_{p,l} & 0 & 0 \\ 0 & 0 & 0 \end{array}\right],$$

$$\mathscr{E}_p(x) = \left[\begin{array}{cc|c} 0 & 0 & E_p x \\ 0 & 0 & \\ \hline (E_p x)^T & & 0 \end{array}\right],$$

$$\mathscr{G}_p(\gamma_p) = \left[\begin{array}{ccc} 0 & 0 & 0 \\ 0 & -\gamma_p & 0 \\ 0 & 0 & -\gamma_p \end{array}\right],$$

and $\zeta_l = -2\cos(l\omega_p \delta T_s)$ and $\eta_l = \exp(jl\omega_p T_s)$.

This way, optimization problem (A.1) is transformed into the following, equivalent SDP:

$$\text{minimize}\quad \gamma_p + \alpha\gamma_{np} \tag{A.6a}$$

$$\text{subject to}\quad \mathscr{K}_{np}(P_{np}) + \mathscr{E}_{np}(x) + \mathscr{G}_{np}(\gamma_{np}) + \mathscr{F}_{np} \preceq 0 \tag{A.6b}$$

$$\mathscr{K}_p(P_{p,l}) + \mathscr{R}_{p,l}(Q_{p,l}) + \mathscr{E}_p(x) + \mathscr{G}_p(\gamma_p) + \mathscr{F}_p \preceq 0, \quad \forall l \in \mathscr{L} \tag{A.6c}$$

$$Q_{p,l} \succeq 0, \qquad\qquad\qquad\qquad\qquad\qquad\quad \forall l \in \mathscr{L}, \tag{A.6d}$$

with optimization variables γ_p, γ_{np}, x, P_{np}, and $P_{p,l}$, $Q_{p,l}$ for all $l \in \mathscr{L}$. The computational complexity of solving this SDP with a standard interior-point solver equals $\mathscr{O}(n^6)$.

A.2.3   Dual SDP

Alternative to solving the primal SDP (A.6), its dual can be solved at the same computational complexity. The formulation of the dual of (A.6) is facilitated by the notion of adjoint mappings: Given a linear mapping $\mathscr{F}$ from vector space $\mathbf{V}$ with internal product $\langle\cdot,\cdot\rangle_{\mathbf{V}}$ to vector space $\mathbf{W}$ with $\langle\cdot,\cdot\rangle_{\mathbf{W}}$, then $\mathscr{F}^{\text{adj}} : \mathbf{W} \to \mathbf{V}$, such that

$$\forall \alpha \in \mathbf{V}, \beta \in \mathbf{W} : \langle\mathscr{F}(\alpha),\beta\rangle_{\mathbf{W}} = \langle\alpha,\mathscr{F}^{\text{adj}}(\beta)\rangle_{\mathbf{V}}.$$

The adjoints of the linear mappings defined in the previous section are given by

$$\mathscr{K}_{np}^{\text{adj}}(Z_{np}) = \begin{bmatrix} A & B_{np} & 0 \end{bmatrix} Z_{np} \begin{bmatrix} A^T \\ B_{np}^T \\ 0 \end{bmatrix} - \begin{bmatrix} I & 0 & 0 \end{bmatrix} Z_{np} \begin{bmatrix} I \\ 0 \\ 0 \end{bmatrix},$$

$$\mathscr{E}_{np}^{\text{adj}}(Z_{np}) = 2E_{np}^T Z_{np}(1:n{+}1, n{+}2),$$

$$\mathscr{G}_{np}^{\text{adj}}(Z_{np}) = \text{Tr}\left\{ Z_{np} \begin{bmatrix} 0 & 0 & 0 \\ 0 & -1 & 0 \\ 0 & 0 & -1 \end{bmatrix} \right\},$$

$$\mathscr{K}_p^{\text{adj}}(Z_{p,l}) = \begin{bmatrix} A & B_p & 0 \end{bmatrix} Z_{p,l} \begin{bmatrix} A^T \\ B_p^T \\ 0 \end{bmatrix} - \begin{bmatrix} I & 0 & 0 \end{bmatrix} Z_{p,l} \begin{bmatrix} I \\ 0 \\ 0 \end{bmatrix},$$

$$\mathcal{R}_{\mathrm{p},l}^{\mathrm{adj}}(Z_{\mathrm{p},l}) = \eta_l \begin{bmatrix} I & 0 & 0 \end{bmatrix} Z_{\mathrm{p},l} \begin{bmatrix} A^T \\ B_{\mathrm{p}}^T \\ 0 \end{bmatrix} + \eta_l^H \begin{bmatrix} A & B_{\mathrm{p}} & 0 \end{bmatrix} Z_{\mathrm{p},l} \begin{bmatrix} I \\ 0 \\ 0 \end{bmatrix} +$$

$$\zeta_l \begin{bmatrix} I & 0 & 0 \end{bmatrix} Z_{\mathrm{p},l} \begin{bmatrix} I \\ 0 \\ 0 \end{bmatrix} ,$$

$$\mathcal{E}_{\mathrm{p}}^{\mathrm{adj}}(Z_{\mathrm{p},l}) = 2 E_{\mathrm{p}}^T \Re\{Z_{\mathrm{p},l}(1:n+1, n+2)\} ,$$

$$\mathcal{G}_{\mathrm{p}}^{\mathrm{adj}}(Z_{\mathrm{p},l}) = \mathrm{Tr}\left\{ Z_{\mathrm{p},l} \begin{bmatrix} 0 & 0 & 0 \\ 0 & -1 & 0 \\ 0 & 0 & -1 \end{bmatrix} \right\} .$$

Using these adjoint mappings, the dual of SDP (A.6) equals

$$\text{maximize} \quad \mathrm{Tr}\{Z_{\mathrm{np}} \mathscr{F}_{\mathrm{np}}\} + \sum_{l \in \mathscr{L}} \mathrm{Tr}\{Z_{\mathrm{p},l} \mathscr{F}_{\mathrm{p}}\} \tag{A.7a}$$

$$\text{subject to} \quad \alpha + \mathcal{G}_{\mathrm{np}}^{\mathrm{adj}}(Z_{\mathrm{np}}) = 0 \tag{A.7b}$$

$$1 + \sum_{l \in \mathscr{L}} \mathcal{G}_{\mathrm{p}}^{\mathrm{ajd}}(Z_{\mathrm{p},l}) = 0 \tag{A.7c}$$

$$\mathcal{E}_{\mathrm{np}}^{\mathrm{adj}}(Z_{\mathrm{np}}) + \sum_{l \in \mathscr{L}} \mathcal{E}_{\mathrm{p}}^{\mathrm{adj}}(Z_{\mathrm{p},l}) = 0 \tag{A.7d}$$

$$\mathcal{K}_{\mathrm{np}}^{\mathrm{adj}}(Z_{\mathrm{np}}) = 0 \tag{A.7e}$$

$$\mathcal{K}_{\mathrm{p}}^{\mathrm{adj}}(Z_{\mathrm{p},l}) = 0 \qquad \forall l \in \mathscr{L} \tag{A.7f}$$

$$\mathcal{R}_{\mathrm{p},l}^{\mathrm{adj}}(Z_{\mathrm{p},l}) \succeq 0 \qquad \forall l \in \mathscr{L} \tag{A.7g}$$

$$Z_{\mathrm{np}} \succeq 0 \tag{A.7h}$$

$$Z_{\mathrm{p},l} \succeq 0 \qquad \forall l \in \mathscr{L} , \tag{A.7i}$$

where $Z_{\mathrm{np}} \in \mathbf{S}_{n+2}$ and $Z_{\mathrm{p},l} \in \mathbf{H}_{n+2}$ for all $l \in \mathscr{L}$, constitute the optimization variables.

A.3   Eliminated Dual SDP

Equality constraints (A.7e) and (A.7f) allow eliminating the largest part of Z_{np} and $Z_{\mathrm{p},l}$ from SDP (A.7), which results in the following reduction of the number of optimization variables:

$$\sharp = \frac{(n+2)(n+3)}{2} + n_{\mathscr{L}}(n+2)^2 \quad \xrightarrow{\text{elimination}} \quad \sharp = (2n+3) + n_{\mathscr{L}}(4n+4) .$$

Hereby, the computational complexity of solving the SDP with an interior-point solver reduces from $\mathcal{O}(n^6)$ to $\mathcal{O}(n^4)$. The central idea of the elimination is explained below while the reader is referred to [150] for more details.

A.3.1 Elimination of Z_{np}

Let $\Theta \in \mathbf{S}_n$, $\kappa \in \mathbf{R}_n$ and $\kappa_{n+1} \in \mathbf{R}$ be defined as follows

$$Z_{np} = \begin{bmatrix} \Theta & \kappa & \star \\ \kappa^T & \kappa_{n+1} & \star \\ \star & \star & \star \end{bmatrix} ,$$

then constraint (A.7e) imposes

$$A\Theta A^T - \Theta + \left(A\kappa B_{np}^T + B_{np}\kappa^T A^T + B_{np}\kappa_{n+1}B_{np}^T \right) = 0 .$$

For given κ and κ_{n+1}, this corresponds to a Lyapunov equation in Θ and since A is stable, this equation has a unique symmetric solution. Hence, constraint (A.7e) relates Θ uniquely to κ, and this relationship is made explicit in the following way:

$$\Theta = \sum_{i=1}^{n+1} \kappa_i \Theta_i , \tag{A.8a}$$

where

$$A\Theta_i A^T - \Theta_i + \left(A e_i B_{np}^T + B_{np} e_i^T A^T \right) = 0 , \qquad \forall i = 1,\ldots,n , \tag{A.8b}$$

$$A\Theta_i A^T - \Theta_i + \left(B_{np} B_{np}^T \right) = 0 , \qquad i = n+1 , \tag{A.8c}$$

and vector e_i corresponds to the i'th unity vector of $\mathbf{R}_n$. Hence, computing the explicit parametrization (A.8) involves solving $n+1$ Lyapunov equations *a priori*, which requires a computational cost of $\mathcal{O}(n^4)$.

A.3.2 Elimination of $Z_{p,l}$

Since A and B_p are real matrices, complex matrix constraint (A.7f) is equivalent to the following real-valued constraints:

$$\Re\left\{ \mathcal{H}_p^{\text{adj}}(Z_{p,l}) \right\} = \mathcal{H}_p^{\text{adj}}\left(\Re\{Z_{p,l}\} \right) = 0 , \tag{A.9a}$$

$$\Im\left\{ \mathcal{H}_p^{\text{adj}}(Z_{p,l}) \right\} = \mathcal{H}_p^{\text{adj}}\left(\Im\{Z_{p,l}\} \right) = 0 . \tag{A.9b}$$

As a result, separate eliminations can be performed for the real and imaginary part of $Z_{p,l}$. The elimination of $\Re\{Z_{p,l}\}$ is similar to the elimination of Z_{np}, described in the previous section. The elimination of $\Im\{Z_{p,l}\}$ is slightly different, since $\Im\{Z_{p,l}\}$ is skew-symmetric instead of symmetric and is further elaborated below.

Let $\rho \in \mathbf{R}_n$ and the skew-symmetric matrix $\Sigma \in \mathbf{R}_{n \times n}$ be defined as follows

$$\Im\{Z_{p,l}\} = \begin{bmatrix} \Sigma & \rho & \star \\ -\rho^T & 0 & \star \\ \star & \star & 0 \end{bmatrix} ,$$

then constraint (A.9b) imposes

$$A\,\Sigma\,A^T - \Sigma + \left(A\rho\,B_p^T - B_p\rho^T A^T\right) = 0 .$$

For a given ρ, this corresponds to a Lyapunov equation in Σ and this equation has a unique skew-symmetric solution, since A is stable. Hence, constraint (A.9b) relates Σ uniquely to ρ, which is made explicit in the following way:

$$\Sigma = \sum_{i=1}^{n} \rho_i \Sigma_i , \tag{A.10a}$$

where

$$A\,\Sigma_i\,A^T - \Sigma_i + \left(A\,e_i\,B_p^T - B_p\,e_i^T\,A^T\right) = 0 , \quad \forall i = 1,\ldots,n . \tag{A.10b}$$

Hence, computing the explicit parametrization (A.10) requires solving n Lyapunov equations *a priori*, which requires a computational cost of $\mathscr{O}(n^4)$.

A.4   Low-rank Structured Primal SDP

Based on the theory of Popov functions, summarized in Section A.4.1 for a discrete-time SISO LTI system, optimization problem (A.1) is reformulated as an SDP that features a low-rank structure. Solver `gkypsdp`, developed by Liu and Vandenberghe [94] exploits this structure and hereby reduces the computational complexity to $\mathscr{O}(n^3)$. Section A.4.2 details the corresponding LMI reformulations for the constraints of (A.1).

A.4.1 Popov Functions

Any function of the form

$$
\begin{bmatrix} (zI-A)^{-1}B \\ 1 \end{bmatrix}^* R \begin{bmatrix} (zI-A)^{-1}B \\ 1 \end{bmatrix} ,
$$

with a given matrix $R \in \mathbf{H}_{n+1}$, is called a Popov function, where superscript $(\cdot)^*$ indicates the following relation

$$
\begin{bmatrix} (zI-A)^{-1}B \\ 1 \end{bmatrix}^* = \begin{bmatrix} (\frac{1}{z}I-A)^{-1}B \\ 1 \end{bmatrix}^T .
$$

For z on the unit circle: $z^{-1} = z^H$, and hence,

$$
\begin{bmatrix} (zI-A)^{-1}B \\ 1 \end{bmatrix}^* = \begin{bmatrix} (zI-A)^{-1}B \\ 1 \end{bmatrix}^H .
$$

The (generalized) KYP lemma analyzes functions of the form

$$
\begin{bmatrix} H(z) \\ 1 \end{bmatrix}^H \Pi \begin{bmatrix} H(z) \\ 1 \end{bmatrix}
$$

on (parts of) the unit circle, or stated otherwise, the (generalized) KYP lemma analyzes the Popov function

$$
\begin{bmatrix} (zI-A)^{-1}B \\ 1 \end{bmatrix}^* R_\Pi \begin{bmatrix} (zI-A)^{-1}B \\ 1 \end{bmatrix} ,
$$

on (parts of) the unit circle, where

$$
R_\Pi = \begin{bmatrix} C & D \\ 0 & 1 \end{bmatrix}^T \Pi \begin{bmatrix} C & D \\ 0 & 1 \end{bmatrix} .
$$

Identically Zero Popov Functions

Hassibi *et al.* [62] show that

$$
\begin{bmatrix} (zI-A)^{-1}B \\ 1 \end{bmatrix}^* R \begin{bmatrix} (zI-A)^{-1}B \\ 1 \end{bmatrix} = 0 , \quad \forall z \in \mathbf{C} \tag{A.11}
$$

holds if and only if there exists a matrix $P \in \mathbf{H}_n$, for which

$$
\begin{bmatrix} A & B \\ I & 0 \end{bmatrix}^T (\Phi \otimes P) \begin{bmatrix} A & B \\ I & 0 \end{bmatrix} = R . \tag{A.12}
$$

Equation A.11 involves a rational function of order $2n$ in z and consequently, this function is identically zero if it is zero at $2n+1$ distinct points. If these interpolation points are chosen on the unit circle and grouped in the set $\mathscr{C}$, constraint (A.11), and hence, linear matrix equality constraint (A.12), is equivalent to

$$\begin{bmatrix} (zI-A)^{-1}B \\ 1 \end{bmatrix}^H R \begin{bmatrix} (zI-A)^{-1}B \\ 1 \end{bmatrix} = 0, \quad \forall z \in \mathscr{C}.$$

KYP Lemma

LMI (A.4), obtained from the KYP lemma, is equivalent to

$$\begin{bmatrix} A & B \\ I & 0 \end{bmatrix}^T (\Phi \otimes P) \begin{bmatrix} A & B \\ I & 0 \end{bmatrix} = Y - R_\Pi, \tag{A.13}$$

for an arbitrary $Y \in \mathbf{S}_{n+1}$, $Y \preceq 0$. Using the results of the previous paragraph, linear matrix equality (A.13) is equivalent to

$$\begin{bmatrix} (zI-A)^{-1}B \\ 1 \end{bmatrix}^H Y \begin{bmatrix} (zI-A)^{-1}B \\ 1 \end{bmatrix} = \begin{bmatrix} H(z) \\ 1 \end{bmatrix}^H \Pi \begin{bmatrix} H(z) \\ 1 \end{bmatrix}, \quad \forall z \in \mathscr{C}. \tag{A.14}$$

Hence, frequency domain inequality (A.3) holds for all $z \in \Upsilon(\Phi,0)$ if and only if there exists $Y \in \mathbf{S}_{n+1}$, $Y \preceq 0$, for which (A.14) holds.

Generalized KYP Lemma

LMI (A.5b), obtained from the generalized KYP lemma, is reformulated as follows

$$\begin{bmatrix} A & B \\ I & 0 \end{bmatrix}^T (\Phi \otimes P) \begin{bmatrix} A & B \\ I & 0 \end{bmatrix} = Y - R_\Pi - \begin{bmatrix} A & B \\ I & 0 \end{bmatrix}^T (\Psi \otimes Q) \begin{bmatrix} A & B \\ I & 0 \end{bmatrix}, \tag{A.15}$$

with arbitrary $Y \in \mathbf{H}_{n+1}$, $Y \preceq 0$. Using the results on identically zero Popov functions, linear matrix equality (A.15) is equivalent to

$$\begin{bmatrix} (zI-A)^{-1}B \\ 1 \end{bmatrix}^H \left(Y - \begin{bmatrix} g(z)Q & 0 \\ 0 & 0 \end{bmatrix} \right) \begin{bmatrix} (zI-A)^{-1}B \\ 1 \end{bmatrix}$$
$$= \begin{bmatrix} H(z) \\ 1 \end{bmatrix}^H \Pi \begin{bmatrix} H(z) \\ 1 \end{bmatrix}, \quad \forall z \in \mathscr{C}, \tag{A.16}$$

where

$$g(z) = \frac{1}{z} \exp(j\omega_c T_s) + z \exp(-j\omega_c T_s) - 2\cos(\omega_d T_s).$$

Hence, frequency domain inequality (A.3) holds for all $z \in \Upsilon(\Phi, \Psi)$ if and only if there exists $Y \in \mathbf{H}_{n+1}$, $Y \preceq 0$, and $Q \in \mathbf{H}_n$, $Q \succeq 0$, for which (A.16) holds.

A.4.2 Reformulated Constraints

By applying the theory of Popov functions detailed in the previous section, constraint (A.1b) is converted into the following set of constraints in $x \in \mathbf{R}_M$ and $Y_{np} \in \mathbf{S}_{n+1}$:

$$\begin{bmatrix} (zI-A)^{-1}B_{np} \\ 1 \end{bmatrix}^H Y_{np} \begin{bmatrix} (zI-A)^{-1}B_{np} \\ 1 \end{bmatrix} = \gamma_{np}, \quad \forall z \in \mathscr{C}_{np},$$

$$\begin{bmatrix} Y_{np} & E_{np}x + F_{np} \\ (E_{np}x + F_{np})^T & \gamma_{np} \end{bmatrix} \succeq 0,$$

where $\mathscr{C}_{np}$ corresponds to a set of $2n+1$ interpolation points on the unit circle. In [94] it is explained how to choose these interpolation points in order to obtain a well-conditioned SDP.

In addition, each of the constraint (A.1c) is equivalent to the following set of constraints in $x \in \mathbf{R}_M$, $Y_{p,l} \in \mathbf{H}_{n+1}$ and $Q_{p,l} \in \mathbf{H}_n$:

$$\begin{bmatrix} (zI-A)^{-1}B_p \\ 1 \end{bmatrix}^H \left(Y_{p,l} + \begin{bmatrix} g_l(z)Q_{p,l} & 0 \\ 0 & 0 \end{bmatrix} \right) \begin{bmatrix} (zI-A)^{-1}B_p \\ 1 \end{bmatrix} = \gamma_p, \quad \forall z \in \mathscr{C}_p,$$

$$\begin{bmatrix} Y_{p,l} & E_p x + F_p \\ (E_p x + F_p)^T & \gamma_p \end{bmatrix} \succeq 0,$$

$$Q_{p,l} \succeq 0,$$

where

$$g_l(z) = \frac{1}{z} \exp(jl\omega_p T_s) + z\exp(-jl\omega_p T_s) - 2\cos(l\omega_p \delta T_s),$$

and $\mathscr{C}_p$ corresponds to a set of $2n+1$ interpolation points on the unit circle.

Appendix B
Introduction to Output Regulation

This appendix provides a brief introduction to output regulation, which concerns the design of an internally stabilizing controller that yields perfect asymptotic tracking/rejection of persistent inputs (see e.g. [117] for an in-depth treatment). Persistent signals have infinite energy and can be described as the autonomous output of a marginally stable system. According to the focus of this monograph, regulation theory is elaborated here for a discrete-time SISO LTI system, while the persistent input is considered periodic and generated by $\Lambda(z)$, defined by Equation 2.5. Hence, output regulation theory is applied to design a controller that yields perfect periodic performance. Since current regulation theory cannot cope with uncertainty on $\Lambda(z)$, period-time uncertainty cannot be accounted for: $\boldsymbol{\delta} = 0\%$.

As the output regulation problem is generally handled in state space, Section B.1 first presents the state-space models of $P(z)$ and $\Lambda(z)$. Subsequently, Section B.2 formulates the Internal Model Principle, the keystone of regulation theory, which states that the controller must contain a (partial) copy of $\Lambda(z)$ to achieve output regulation. The design of the remaining freedom in the controller is facilitated by the derivation of an auxiliary plant, detailed in Section B.3, while Section B.4 presents particular controllers that achieve output regulation. To conclude, Section B.5 discusses the modifications required to relax the perfect periodic performance enforced by the Internal Model Principle. To alleviate notation, argument (k) of the sampled time signals is temporarily omitted.

B.1 Control Problem Formulation

This appendix follows the control problem formulation of Section 2.2, but requires a state-space model of $P(z)$ and $\Lambda(z)$. These models are first presented below, followed by the formulation of the output regulation problem.

State-Space Models of $P(z)$ and $\Lambda(z)$

The state-space model of generalized plant $P(z)$ is described as follows:

$$P(z) : \begin{cases} qx = Ax + \sum_{i \in \mathscr{I}} B_i w_i + B_u u \\ v_i = C_i x + D_i w_i + D_{iu} u \\ y = C_y x + \sum_{i \in \mathscr{I}} D_{yi} w_i \end{cases} . \tag{B.1}$$

Index $i \in \mathscr{I}$ labels the design specifications in the controller design, each involving the closed-loop subsystem from exogenous input w_i to regulated outputs v_i. As the off-diagonal subsystems are irrelevant to the controller design, only the direct feed-through terms from w_i to v_i are indicated in (B.1). Without loss of generality, see e.g. [78], no direct feed-through term from u to y is considered.

Perfect periodic performance corresponds to the i_p'th design specification, where subscript $(\cdot)_{i_p}$ is shortened to $(\cdot)_p$. Periodic input w_p corresponds to the autonomous output of signal generator $\Lambda(z)$, Equation 2.5, described by the following state-space model:

$$\Lambda(z) : \begin{cases} qx_\Lambda = A_\Lambda x_\Lambda \\ w_p = C_\Lambda x_\Lambda \end{cases} . \tag{B.2}$$

In regulation theory, input w_p is replaced by the states x_Λ of its signal generator, and to shorten notation, B_p, D_p and D_{yp} are redefined as follows:

$$B_p \equiv B_p C_\Lambda ,$$
$$D_p \equiv D_p C_\Lambda ,$$
$$B_{yp} \equiv D_{yp} C_\Lambda .$$

The remainder of this appendix focusses on one additional performance specification i besides i_p, and continues with the following state-space model:

$$P(z) : \begin{cases} qx = Ax + B_i w_i + B_p x_\Lambda + B_u u \\ v_i = C_i x + D_i w_i + D_{iu} u \\ v_p = C_p x + D_p x_\Lambda + D_{pu} u \\ y = C_y x + D_{yi} w_i + D_{yp} x_\Lambda \end{cases} .$$

Output Regulation

Output regulation deals with the design of a feedback controller $K(z)$ that internally stabilizes the closed-loop system and guarantees

$$\lim_{k \to \infty} v_p(k) = 0 , \tag{B.3}$$

independent of the initial states of $P(z)$, $\Lambda(z)$ and $K(z)$.

Well-posedness of this control problem requires the following assumptions:

1. The pair (A, B_u) is stabilizable.
2. The pair $\left(\begin{bmatrix} C_y & D_{yp} \end{bmatrix}, \begin{bmatrix} A & B_p \\ 0 & A_\Lambda \end{bmatrix} \right)$ is detectable.

The combination of (A, B_u) stabilizable and (C_y, A) detectable is essential to guarantee the existence of an internally stabilizing controller, while $\Lambda(z)$ can always be reduced to satisfy assumption 2, without loss of generality [45].

B.2   Internal Model Principle

As proven in e.g. [117], the existence of a controller that achieves output regulation is equivalent to the existence of matrices Π and Γ that solve the following linear equation, often called the regulator equation:

$$\Pi A_\Lambda = A\Pi + B_u \Gamma + B_p , \tag{B.4a}$$

$$0 = C_p \Pi + D_{pu} \Gamma + D_p . \tag{B.4b}$$

For a SISO control problem, the regulator equation has a solution if and only if subsystem $P_{pu}(z)$, from u to v_p, has no zeros coinciding with an eigenvalue of A_Λ.

The Internal Model Principle states that a controller achieves output regulation if and only if it admits a realization of the form:

$$K(z) : \begin{cases} q\xi_1 = A_\Lambda \xi_1 + C_{K1}\xi_2 + D_{K1}\left(y + (D_{yp} + C_y\Pi)\xi_1\right) \\ q\xi_2 = A_K \xi_2 + B_K \left(y + (D_{yp} + C_y\Pi)\xi_1\right) \\ u = -\Gamma \xi_1 + C_{K2}\xi_2 + D_{K2}\left(y + (D_{yp} + C_y\Pi)\xi_1\right) \end{cases} . \tag{B.5}$$

Matrices Π and Γ are the solution of (B.4), while the design of A_K, B_K, C_{K1}, C_{K2}, D_{K1} and D_{K2} is free as long as it guarantees internal closed-loop stability.

The earliest and most well-known form of the Internal Model Principle concerns the control problem where $v_p(k)$ corresponds to the measured output: $v_p(k) = y(k)$. By this equality, (B.4b) yields $D_{yp} + C_y\Pi = 0$, such that $K(z)$ obtains a state matrix of the following form:

$$\begin{bmatrix} A_\Lambda & C_{K1} \\ 0 & A_K \end{bmatrix} ,$$

and hence, the poles of $\Lambda(z)$ are reproduced in $K(z)$.

While controller structure (B.5) guarantees perfect periodic performance, the design of A_K, B_K, C_{K1}, C_{K2}, D_{K1} and D_{K2} should address the remaining design specifications $i \in \mathscr{I} \setminus \{i_p\}$. Since designing a controller with a particular structure is mathematically involved, it is more convenient to transfer states ξ_1, which relate to the output regulation requirement, to the plant. This yields an auxiliary plant $\overline{P}(z)$, and designing an unstructured controller $\overline{K}(z)$ for $\overline{P}(z)$ is equivalent to designing a structured controller $K(z)$, Equation B.5, for $P(z)$, see e.g. [117] for a proof.

B.3 Auxiliary Plant

Transferring states ξ_1 from controller $K(z)$ to plant $P(z)$ requires the following modifications:

- The controller is provided with an additional output u_1 that transfers the dynamics of ξ_1 to the plant: $u_1 = q\xi_1 - A_\Lambda \xi_1$.
- Control signal u is replaced by $u_2 = u + \Gamma \xi_1$.
- Output signal y is replaced by $\bar{y} = y + (D_{yp} + C_y \Pi)\xi_1$.

The remaining controller dynamics are contained in $\overline{K}(z)$:

$$\overline{K}(z) : \begin{cases} q\xi_2 = A_K \xi_2 + B_K \bar{y} \\ u_1 = C_{K1}\xi_2 + D_{K1}\bar{y} \\ u_2 = C_{K2}\xi_2 + D_{K2}\bar{y} \end{cases},$$

while the resulting auxiliary plant $\overline{P}(z)$ is given by

$$\overline{P}(z) : \begin{cases} qx = Ax + B_i w_i + B_p x_\Lambda + B_u(u_2 - \Gamma\xi_1) \\ q\xi_1 = A_\Lambda \xi_1 + u_1 \\ v_i = C_i x + D_i w_i + D_{iu}(u_2 - \Gamma\xi_1) \\ v_p = C_p x + D_p x_\Lambda + D_{pu}(u_2 - \Gamma\xi_1) \\ \bar{y} = C_y x + D_{yi} w_i + D_{yp} x_\Lambda + (D_{yp} + C_y \Pi)\xi_1 \end{cases}.$$

More insight in the auxiliary plant dynamics is obtained by the following state transformation:

$$x_1 = x - \Pi x_\lambda ,$$
$$x_2 = \xi_1 + x_\lambda ,$$

which yields:

$$\overline{P}(z) : \begin{cases} qx_1 = Ax_1 - B_u \Gamma x_2 + B_i w_i + B_u u_2 \\ qx_2 = + A_\Lambda x_2 + u_1 \\ v_i = C_i x_1 - D_{iu}\Gamma x_2 + D_i w_i + D_{iu} u_2 + (C_i \Pi + D_{iu}\Gamma)x_\Lambda \\ v_p = C_p x_1 - D_{pu}\Gamma x_2 + D_{pu} u_2 \\ \bar{y} = C_y x_1 + (D_{yp} + C_y \Pi)x_2 + D_{yi} w_i \end{cases}.$$

The direct feed-through term from x_Λ to v_i can be omitted as it is irrelevant to the controller design. As long as $\overline{K}(z)$ yields an internally stable closed-loop system, output regulation is always obtained, since in $\overline{P}(z)$ neither the state equation, nor the output equations for v_p and $\bar{y}$ depend on x_Λ.

B.4   Special Regulators

If both x and x_Λ are measurable (if the pair (C_Λ, A_Λ) is observable, x_Λ can be re-constructed from the n_Λ last samples of w_p), a static state-feedback controller of the form

$$u = \begin{bmatrix} (\Gamma - F\Pi) & F \end{bmatrix} \begin{bmatrix} x_\Lambda \\ x \end{bmatrix} , \tag{B.6}$$

where F is an arbitrary matrix such that $A + B_u F$ is stable, achieves output regulation.

For measurement feedback control, regulation is achieved by a controller of the form

$$K(z) : \begin{cases} q \begin{bmatrix} \hat{x}_\Lambda \\ \hat{x} \end{bmatrix} = \begin{bmatrix} A_\Lambda & 0 \\ B_p & A \end{bmatrix} \begin{bmatrix} \hat{x}_\Lambda \\ \hat{x} \end{bmatrix} + \begin{bmatrix} 0 \\ B_u \end{bmatrix} u(k) + L \left(\begin{bmatrix} D_{yp} & C_y \end{bmatrix} \begin{bmatrix} \hat{x}_\Lambda \\ \hat{x} \end{bmatrix} - y \right) \\ u = \begin{bmatrix} (\Gamma - F\Pi) & F \end{bmatrix} \begin{bmatrix} \hat{x}_\Lambda \\ \hat{x} \end{bmatrix} \end{cases}$$

where F and L are chosen such that

$$A + B_u F \quad \text{and} \quad \begin{bmatrix} A_\Lambda & 0 \\ B_p & A \end{bmatrix} + L \begin{bmatrix} D_{yp} & C_y \end{bmatrix}$$

are stable. This controller corresponds to the combination of a Luenberger state observer [96] and state-feedback controller (B.6).

B.5   Extension to Approximate Regulation

Only few contributions in output regulation deal with approximate instead of perfect regulation (B.3) [72, 86, 88, 93]. Köroğlu and Scherer [88] present an elegant relaxation of exact output regulation theory, where "almost asymptotic regulation of level κ" involves a controller design that internally stabilizes the closed-loop system and guarantees

$$\|v_p(k)\|_2 \le \kappa \|x_\Lambda(k)\|_2 ,$$

for $k \to \infty$, independent of the initial states of $P(z)$, $\Lambda(z)$ and $K(z)$. State-space model (B.2) of $\Lambda(z)$ can be chosen such that $\|x_\Lambda(k)\|_2$ equals $\mathrm{rms}(w_p(k))$. The controller design for almost output regulation is very similar to the design for exact regulation, where the main modification is replacing Equation B.4b by:

$$\begin{bmatrix} \kappa I & (C_p\Pi + D_{pu}\Gamma + D_p) \\ (C_p\Pi + D_{pu}\Gamma + D_p)^T & \kappa I \end{bmatrix} \succeq 0 .$$

Appendix C
Robust Controller Design Using the Structured Singular Value

Section 3.3.4 presents an intuitive robust feedforward controller design approach, but it only applies to the control configurations of Figures 3.1(a) and 3.1(b). Alternatively, Section 4.3.4 proposes a robust disturbance feedback controller design, but it cannot handle the robust tracking requirement (4.11) in an intuitive way. This appendix presents the more general approach to the robust feedforward and estimated disturbance feedback controller design, which relies on the structured singular value, see e.g. [43, 106, 131] for more details.

First, Section C.1 presents the definition and some properties of the structured singular value. Subsequently, Section C.2 tackles the robust feedforward controller design using the structured singular value, while Section C.3 deals with the robust disturbance feedback controller design.

C.1 Structured Singular Value

While this monograph generally considers uncertainty set (2.2b), this section considers a more general set $\boldsymbol{\Delta}$, which allows for MIMO plant uncertainty $\Delta(z)$. The structure of this uncertainty is determined by subset $\boldsymbol{\Psi} \subset \mathbf{C}_{m \times m}$:

$$\boldsymbol{\Psi} = \left\{ \mathrm{diag}(\psi_1 I_{m_1}, \ldots \psi_E I_{m_E}, \Psi_1, \ldots, \Psi_F) \mid \psi_i \in \mathbf{C}, \, \Psi_j \in \mathbf{C}_{m_j \times m_j} \right\}, \qquad (C.1)$$

and the more general set $\boldsymbol{\Delta}$ considered in this section is

$$\boldsymbol{\Delta} = \left\{ \Delta(z) \text{ is a stable system with } \|\Delta(z)\|_\infty \leq 1, \text{ and } \forall z \in \mathbf{C} : \Delta(z) \in \boldsymbol{\Psi} \right\}. \qquad (C.2)$$

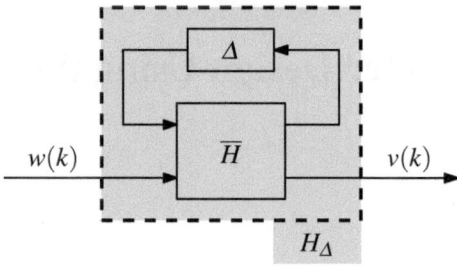

Fig. C.1 Uncertain system H_Δ corresponding to the linear fractional transformation of $\overline{H}$ and Δ.

Definition

For a given matrix $M \in \mathbf{C}_{m \times m}$, the structured singular value related to $\mathbf{\Psi}$ (C.1), denoted by $\mu_{\mathbf{\Psi}}(M)$, is defined as:

$$\mu_{\mathbf{\Psi}}(M) \equiv \frac{1}{\min\{\sigma_{\max}(\Psi) \mid \Psi \in \mathbf{\Psi}, \, \det(I - M\Psi) = 0\}},$$

unless no $\Psi \in \mathbf{\Psi}$ makes $(I - M\Psi)$ singular, in which case $\mu_{\mathbf{\Psi}}(M) \equiv 0$. $\sigma_{\max}(\Psi)$ denotes the largest singular value of matrix Ψ.

Computation

In general, the structured singular value $\mu_{\mathbf{\Psi}}(M)$ cannot be computed analytically, and its numerical computation involves a nonconvex optimization problem. On the other hand, the following upper-bound on $\mu_{\mathbf{\Psi}}(M)$ can be computed using convex optimization:

$$\mu_{\mathbf{\Psi}}(M) \leq \min\{\sigma_{\max}(PMP^{-1}) \mid P \in \mathbf{P}\}, \tag{C.3}$$

where $\mathbf{P}$ is the set of matrices P that commute with all $\Psi \in \mathbf{\Psi}$, i.e., satisfy $P\Psi = \Psi P$. The resulting upper-bound is generally tight (within a few per cent), and even exact if $2E + F \leq 3$. Using upper bound (C.3), $\mu_{\mathbf{\Delta}}(M) < 1$ is guaranteed if there exists a matrix $Q = P^{-1}P^{-H}$, where $P \in \mathbf{P}$, that satisfies:

$$MQM^H - Q \prec 0.$$

Application to Robust Performance Analysis

Figure C.1 shows an uncertain (closed-loop) system $H_\Delta(z)$, where model uncertainty $\Delta(z) \in \mathbf{\Delta}$ (C.2) is "pulled-out", such that $H_\Delta(z)$ corresponds to the linear fractional transformation of $\overline{H}(z)$ and $\Delta(z)$. At a given frequency ω, robust closed-loop performance can be analyzed by means of the following equivalence:

(a)

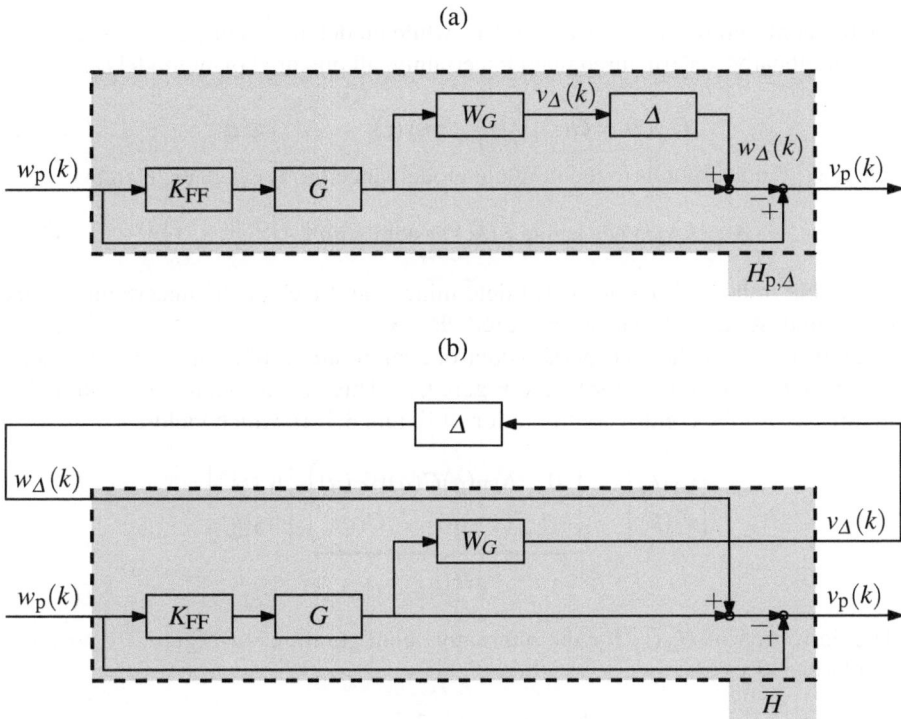

(b)

Fig. C.2 Transformation of the robust feedforward control configuration (a) to the linear fractional representation (b), where dynamic uncertainty Δ is "pulled-out" by means of additional input $w_\Delta(k)$ and output $v_\Delta(k)$.

$$|H_\Delta(\omega)| < 1, \ \forall \Delta(z) \in \pmb{\Delta} \quad \Leftrightarrow \quad \mu_{\overline{\pmb{\Psi}}}\big(\overline{H}(\omega)\big) < 1,$$

where

$$\overline{\pmb{\Psi}} = \big\{ \mathrm{diag}(\pmb{\Psi}_1, \pmb{\Psi}_2) \mid \pmb{\Psi}_1 \in \pmb{\Psi}, \ \pmb{\Psi}_2 \in \mathbf{C}_{m_w \times m_v} \big\}.$$

m_w and m_v respectively denote the dimensions of exogenous input $w(k)$ and regulated output $v(k)$.

C.2 Robust Feedforward Controller Design

The difficulty in the robust feedforward controller design is to render the following constraint:

$$|H_{\mathrm{p},\Delta}(\omega)| \le \nu, \quad \forall \Delta(z) \in \pmb{\Delta}, \tag{C.4}$$

for given ω, convex in ν and the design parameters of feedforward controller $K_{\mathrm{FF}}(z)$. The relation between closed-loop system $H_{\mathrm{p},\Delta}(z)$ and $K_{\mathrm{FF}}(z)$ is determined by the control configuration (see Section 3.2.1), while model uncertainty $\Delta(z)$ corresponds to multiplicative unstructured plant uncertainty: all potential plant models are of the form

$$G_\Delta(z) = G(z)[1 + W_G(z)\Delta(z)] , \quad \Delta(z) \in \Delta , \tag{C.5a}$$

where $G(z)$ corresponds to the nominal model, uncertainty set Δ is given by

$$\Delta = \{\Delta(z) \text{ is a stable SISO system with } \|\Delta(z)\|_\infty \leq 1\} , \tag{C.5b}$$

and stable transfer function $W_G(z)$ determines the "size" of the uncertainty. Since uncertainty set Δ (C.5b) is unstructured: $\Psi = C$.

In the first step, the feedforward control configurations of Figure 3.1 with uncertain plant $G_\Delta(z)$ are transformed to Figure C.1. This transformation is illustrated in Figure C.2 for the control configuration of Figure 3.1(a), which yields:

$$\begin{bmatrix} v_\Delta(k) \\ v_p(k) \end{bmatrix} = \underbrace{\begin{bmatrix} 0 & K_{FF}(q)G(q)W_G(q) \\ -1 & (1 - K_{FF}(q)G(q)) \end{bmatrix}}_{\overline{H}(q)} \begin{bmatrix} w_\Delta(k) \\ w_p(k) \end{bmatrix} .$$

The derivation of $H_\Delta(z)$ for the alternative configurations of Figure 3.1 proceeds similarly and yields:

$$(b) : \overline{H}(z) = \begin{bmatrix} 0 & K_{FF}GW_G \\ 1 & G_d + K_{FF}G \end{bmatrix} ,$$

$$(c) : \overline{H}(z) = \begin{bmatrix} -T_oW_G & K_{FF}T_oW_G \\ -S_o & 1 - K_{FF}T_o \end{bmatrix} ,$$

$$(d) : \overline{H}(z) = \begin{bmatrix} -T_oW_G & S_o(K_{FF} + K_o)GW_G \\ -S_o & S_o(1 - K_{FF}G) \end{bmatrix} ,$$

$$(e) : \overline{H}(z) = \begin{bmatrix} -T_oW_G & S_o(K_{FF} - G_dK_o)GW_G \\ S_o & S_o(G_d + K_{FF}G) \end{bmatrix} ,$$

where in the right-hand side, argument (z) is omitted to save space. To conclude, all configurations give rise to

$$\overline{H}(z) = \begin{bmatrix} H_1(z) & H_{1p}(z) \\ H_{p1}(z) & H_p(z) \end{bmatrix} ,$$

where $H_1(z)$ and $H_{p1}(z)$ are independent of $K_{FF}(z)$, while $H_{1p}(z)$ and $H_p(z)$ depend affinely on $K_{FF}(z)$.

In the second step, constraint (C.4) is analyzed by means of the structured singular value. In this case, $\overline{\Psi}$ is given by

$$\overline{\Psi} = \{\text{diag}(\Psi_1, \Psi_2) \mid \Psi_1 \in C, \Psi_2 \in C\} , \tag{C.6}$$

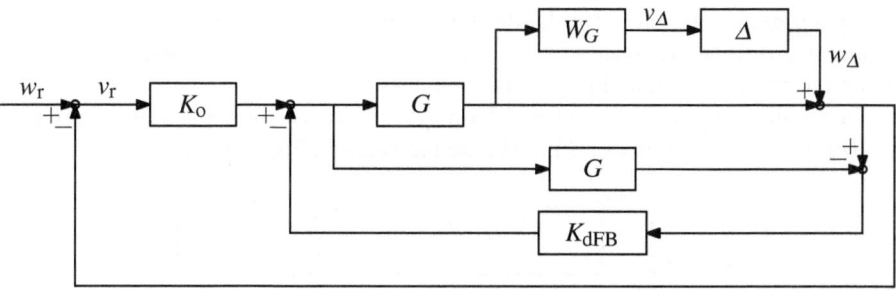

Fig. C.3 Robust estimated disturbance feedback control configuration, where argument (k) of the sampled time signals is omitted to save space.

and the set $\mathbf{P}$ of matrices that commute with $\Psi \in \overline{\Psi}$ corresponds to $\mathbf{P} = \overline{\Psi}$. This way, all matrices $Q = P^{-1}P^{-H}$, where $P \in \mathbf{P}$, are real and diagonal, and without loss of generality one of the diagonal elements is set equal to 1:

$$Q = \begin{bmatrix} q & 0 \\ 0 & 1 \end{bmatrix}. \tag{C.7}$$

Since $\overline{\Psi}$, Equation C.6, only contains $F = 2$ subblocks, upperbound (C.3) is exact, and therefore, the following equivalences hold:

$$|H_{p,\Delta}(\omega)| \leq v, \quad \forall \Delta(z) \in \Delta$$

$$\Leftrightarrow \begin{bmatrix} H_1(\omega) & \frac{1}{v}H_{1p}(\omega) \\ H_{p1}(\omega) & \frac{1}{v}H_p(\omega) \end{bmatrix} \begin{bmatrix} q & 0 \\ 0 & 1 \end{bmatrix} \begin{bmatrix} H_1(\omega) & \frac{1}{v}H_{1p}(\omega) \\ H_{p1}(\omega) & \frac{1}{v}H_p(\omega) \end{bmatrix}^H - \begin{bmatrix} q & 0 \\ 0 & 1 \end{bmatrix} \prec 0$$

$$\Leftrightarrow \begin{bmatrix} \begin{bmatrix} H_1(\omega) \\ H_{p1}(\omega) \end{bmatrix} \bar{q} \begin{bmatrix} H_1(\omega) \\ H_{p1}(\omega) \end{bmatrix}^H - \begin{bmatrix} \bar{q} & 0 \\ 0 & v \end{bmatrix} & \begin{bmatrix} H_{1p}(\omega) \\ H_p(\omega) \end{bmatrix} \\ \begin{bmatrix} H_{1p}(\omega)^H & H_p(\omega)^H \end{bmatrix} & -v \end{bmatrix} \prec 0.$$

The last matrix inequality is affine in v, $\bar{q} = vq$ and in $H_{1p}(\omega)$ and $H_p(\omega)$, which depend affinely on $K_{FF}(\omega)$, and hence, on the design parameters of $K_{FF}(z)$.

C.3   Robust Estimated Disturbance Feedback Controller Design

The difficulty in the robust estimated disturbance feedback controller design is to render the following constraint:

$$|H_{r,\Delta}(\omega)| \leq v, \quad \forall \Delta(z) \in \Delta, \tag{C.8}$$

for given ω and ν, convex in the design parameters of disturbance feedback controller $K_{\text{dFB}}(z)$. Model uncertainty $\Delta(z)$ corresponds to multiplicative unstructured plant uncertainty (C.5) and $H_{r,\Delta}(z)$ is given by (4.11b).

In the first step, auxiliary closed-loop system $\overline{H}(z)$ is derived, where its inputs $w_\Delta(k)$, $w_r(k)$ and outputs $v_\Delta(k)$, $v_r(k)$ are indicated in Figure C.3:

$$\begin{bmatrix} v_\Delta(k) \\ v_r(k) \end{bmatrix} = \underbrace{\begin{bmatrix} -(K_{\text{dFB}} + K_{\text{o}})S_{\text{o}}GW_G & T_{\text{o}}W_G \\ (K_{\text{dFB}}G - 1)S_{\text{o}} & S_{\text{o}} \end{bmatrix}}_{\overline{H}(q)} \begin{bmatrix} w_\Delta(k) \\ w_r(k) \end{bmatrix},$$

where argument (q) is omitted to save space. Using the notation

$$\overline{H}(z) = \begin{bmatrix} H_1(z) & H_{1r}(z) \\ H_{r1}(z) & H_r(z) \end{bmatrix},$$

$H_{1r}(z)$ and $H_r(z)$ are independent of $K_{\text{dFB}}(z)$, while $H_1(z)$ and $H_{r1}(z)$ depend affinely on $K_{\text{dFB}}(z)$.

Second, constraint (C.8) is analyzed by means of the structured singular value, and this step proceeds similarly to the previous section. The only modification required is replacing (C.7) by

$$Q = \begin{bmatrix} 1 & 0 \\ 0 & q \end{bmatrix},$$

whereby the following equivalence is obtained:

$$|H_{r,\Delta}(\omega)| \le \nu, \quad \forall \Delta(z) \in \Delta$$

$$\Leftrightarrow \begin{bmatrix} \begin{bmatrix} H_{1r}(\omega) \\ H_r(\omega) \end{bmatrix} \dfrac{q}{\nu^2} \begin{bmatrix} H_{1r}(\omega) \\ H_r(\omega) \end{bmatrix}^H - \begin{bmatrix} 1 & 0 \\ 0 & q \end{bmatrix} & \begin{bmatrix} H_1(\omega) \\ H_{r1}(\omega) \end{bmatrix} \\ \begin{bmatrix} H_1(\omega)^H & H_{r1}(\omega)^H \end{bmatrix} & -1 \end{bmatrix} \prec 0.$$

References

1. Abedor, J., Nagpal, K., Khargonekar, P.P., Poolla, K.: Robust regulation with an $\mathcal{H}_\infty$ contraint. In: Proc. of the International Workshop on Robust Control; in Control of Uncertain Dynamic Systems, San Antonio, TX, USA, March 1991, pp. 95–110 (1991)
2. Abedor, J., Nagpal, K., Khargonekar, P.P., Poolla, K.: Robust regulation in the presence of norm-bounded uncertainty. IEEE Transactions on Automatic Control 40(1), 147–153 (1995)
3. Abedor, J., Nagpal, K., Poolla, K.: Does robust regulation compromise $\mathcal{H}_2$ performance. In: Proc. of the 31st IEEE Conference on Decision and Control, Tucson, AZ, USA, December 1992, pp. 2002–2007 (1992)
4. Abedor, J., Nagpal, K., Poolla, K.: Robust regulation with $\mathcal{H}_2$ performance. Systems & Control Letters 23(6), 431–443 (1994)
5. Al-Bender, F.: On the modelling of the dynamic characteristics of aerostatic bearing films: from stability analysis to active compensation. Precision Engineering 33(2), 117–126 (2009)
6. Al-Bender, F., Van Brussel, H.: Development of high frequency electrospindles with passive and active air bearings. In: Proc. of the International Seminar on Improving Machine Tool Performance, San Sebastian, Spain, July 1998, pp. 175–184 (1998)
7. Al-Bender, F., Van Brussel, H., Vanherck, P.: Actively compensated aerostatic bearings. In: Proc. of the ASPE 2001, Precision Bearings and Spindles, Summer Tropical Meeting, June 2001, Penn State University, University Park (2001)
8. Åström, K.J., Hagander, P., Sternby, J.: Zeros of sampled systems. Automatica 20(1), 31–38 (1984)
9. Bask, M.P., Medvedev, A.: External model repetitive controller for active vibration isolation. In: Proc. of the 1999 International Symposium on Active Control of Sound and Vibration, Fort Lauderdale, Florida, USA, December 1999, pp. 177–188 (1999)
10. Bode, H.W.: Network Analysis and Feedback Amplifier Design. D. Van Nostrad Company, New York (1945)
11. Bodson, M., Jensen, J.S., Douglas, S.C.: Active noise control for periodic disturbances. IEEE Transactions on Control Systems Technology 9(1), 200–205 (2001)
12. Botteron, F., Pinheiro, H.: Discrete-time internal model controller for three-phase PWM inverters with insulator transformer. IEE Proceedings - Electric Power Applications 153(1), 57–67 (2006)
13. Boyd, S., Barratt, C.: Linear Controller Design: Limits of Performance. Prentice-Hall, Englewood Cliffs (1991)

14. Boyd, S., El Ghaoui, L., Feron, E., Balakrishnan, V.: Linear Matrix Inequalities in Systems and Control Theory. SIAM Studies in Applied Mathematics, vol. 15. Society for Industrial and Applied Mathematics (SIAM), Philadelphia (1994)

15. Boyd, S., Vandenberghe, L.: Convex Optimization. Cambridge University Press, Cambridge (2004)

16. Broberg, H.L., Molyet, R.G.: Correction of periodic errors in a weather satellite servo using repetitive control. In: First IEEE Conference on Control Applications, Dayton, OH, September 1992, pp. 682–683 (1992)

17. Bryan, J.B., Vanherck, P.: Unification of terminology concerning the error motion of axes of rotation. CIRP Annals – Manufacturing Technology 24(2), 555–562 (1975)

18. Butterworth, J.A., Pao, L.Y., Abramovitch, D.Y.: The effect of nonmimimum-phase zero locations on the performance of feedforward model-inverse control techniques in discrete-time systems. In: Proc. of the 2008 American Control Conference, Seattle, WA, USA, June 2008, pp. 2696–2702 (2008)

19. Cevik, M.K.K., Schumacher, J.M.: The robust regulation problem with robust stability: a subspace-valued function approach. IEEE Transactions on Autmomatic Control 45(9), 1735–1738 (2000)

20. Chang, W.S., Suh, I.H., Kim, T.W.: Analysis and design of two types of digital repetitive control systems. Automatica 31(5), 741–746 (1995)

21. Chen, J.: Multivariable gain-phase and sensitivity integral relations and design trade-offs. IEEE Transactions on Automatic Control 43(3), 373–385 (1998)

22. Chen, J.: Logarithmic integrals, interpolation bounds, and performance limitations in MIMO feedback systems. IEEE Transactions on Automatic Control 45(6), 1098–1115 (2000)

23. Chen, X., Zhai, G., Fukuda, T.: An approximate inverse system for nonminimum-phase systems and its application to disturbance observer. Systems & Control Letters 52, 193–207 (2004)

24. Chen, Y.Q., Ding, M.Z., Xiu, L.C., Ooi, K.K., Tan, L.L.: Optimally designed parsimoneous repetitive learning compensator for hard disc drives having high track density (2001) US patent 2001/0043427 A1

25. Chen, Y.Q., Moore, K.L., Yu, J., Zhang, T.: Iterative learning control and repetitive control in hard disk drive industry - a tutorial. In: Proc. of the 45th IEEE Conference on Decision and Control, San Diego, CA, USA, December 2006, pp. 6778–6791 (2006)

26. Chew, K.K., Tomizuka, M.: Digital control of repetitive errors in disk drive systems. In: Proc. of the 1989 American Control Conference, Pittsburgh, PA, USA, June 1989, pp. 540–548 (1989)

27. Chew, K.K., Tomizuka, M.: Digital control of repetitive errors in disc drive systems. IEEE Control Systems Magazine 10(1), 16–20 (1990)

28. Chew, K.K., Tomizuka, M.: Steady-state and stochastic performance of a modified discrete-time prototype repetitive controller. Transactions of the ASME: Journal of Dynamic Systems, Measurement, and Control 112(1), 35–41 (1990)

29. Choi, B.K., Choi, C.H.: An effective pole-zero cancellation in feedforward controllers for nonminimum phase systems. In: Proc. of the 1997 Amerian Control Conference, Albuquerque, New Mexico, June 1997, pp. 3907–3908 (1997)

30. Choi, G.H., Oh, J.H., Choi, G.S.: Repetitive tracking control of a coarse-fine actuator. In: Proc. of the IEEE/ASME International Conference on Advanced Intelligent Mechatronics, Atlanta, USA, September 1999, pp. 335–340 (1999)

31. Choi, Y., Chung, W.K., Youm, Y.: Disturbance observer in $\mathcal{H}_\infty$ frameworks. In: Proc. of the 22nd International Conference on Industrial Electronics, Control, and Instrumentation, Taipei, Taiwan, August 1996, pp. 1394–1400 (1996)

32. Choi, Y., Yang, K., Chung, W.K., Kim, H.R., Suh, I.H.: On the robustness and performance of disturbance observers for second-order systems. IEEE Transactions on Automatic Control 48(2), 315–320 (2003)

33. Cuiyan, L., Dongchun, Z., Xianyi, Z.: A survey of repetitive control. In: Proc. of 2004 IEEE/RSJ International Conference on Intelligent Robots and Systems, Sendai, Japan, September-October 2004, pp. 1160–1166 (2004)

34. Davison, E.J.: Output control of linear time-invariant multivariable systems with unmeasurable arbitrary disturbances. IEEE Transactions on Automatic Control 17(5), 621–630 (1972)

35. Davison, E.J., Goldenberg, A.: The robust control of a general servomechanism problem: the servo compensator. Automatica 11(5), 461–471 (1975)

36. de Oliveira, M.C., Geromel, J.C., Bernussou, J.: Extended $\mathcal{H}_2$ and $\mathcal{H}_\infty$ norm characterizations and controller parametrizations for discrete-time systems. International Journal of Control 75(9), 666–679 (2002)

37. Desoer, C.A., Liu, R.W., Murray, J., Saeks, R.: Feedback system design: the fractional representation approach to analysis and synthesis. IEEE Transactions on Automatic Control 25(3), 399–412 (1980)

38. Devasia, S.: Should model-based inverse inputs be used as feedforward under plant uncertainty. IEEE Transactions on Automatic Control 47(11), 1865–1871 (2002)

39. Devasia, S., Chen, D., Paden, B.: Nonlinear inversion-based output tracking. IEEE Transactions on Automatic Control 41(7), 930–942 (1996)

40. Doh, T.Y., Ryoo, J.R., Chung, M.J.: Design of a repetitive controller: an application to the track-following servo system of optical disk drives. IEE Proceedings - Control Theory and Applications 153(3), 323–330 (2006)

41. Donaldson, R.R.: A simple method for separating spindle error from test ball roundness error. CIRP Annals – Manufacturing Technology 21(1), 125–126 (1972)

42. Dötsch, H.G.M., Smakman, H.T., Van den Hof, P.M.J., Steinbuch, M.: Adaptive repetitive control of a compact disk mechanism. In: Proc. of the 1995 IEEE Conference on Decision and Control, New Orleans, LA, USA, December 1995, pp. 1720–1725 (1995)

43. Doyle, J.: Analysis of feedback systems with structured uncertainty. IEE Proceedings–D, Control Theory and Applications 129(6), 242–250 (1982)

44. Ferreres, G., Roos, C.: Efficient convex design of robust feedforward controllers. In: Proc. of the 44th IEEE Conference on Decision and Control and the European Conference on Control, Seville, Spain, December 2005, pp. 6460–6465 (2005)

45. Francis, B.A.: The linear multivariable regulator problem. SIAM Journal on Control and Optimization 15(3), 486–505 (1977)

46. Francis, B.A., Sebakhy, O.A., Wonham, W.M.: Synthesis of multivarible regulators: the internal model principle. Applied Mathematics and Optimization 1(1), 64–86 (1974)

47. Francis, B.A., Wonham, W.M.: The internal model principle for linear multivariable regulators. Applied Mathematics and Optimization 2(2), 170–194 (1975)

48. Francis, B.A., Wonham, W.M.: The internal model principle of control theory. Automatica 12(5), 457–465 (1976)

49. Freudenberg, J.S., Looze, D.P.: Right half plane poles and zeros and design tradeoffs in feedback systems. IEEE Transactions on Automatic Control 30(6), 555–565 (1985)

50. Gangloff, J., Ginhoux, R., de Mathelin, M., Soler, L., Marescaux, J.: Model predictive control for compensation of cyclic organ motions in teleoperated laparoscopic surgery. IEEE Transactions on Control Systems Technology 14(2), 235–246 (2006)

51. Gibbs, J.W.: Letter to the editor. Nature 59, 606 (1898)

52. Giusto, A., Paganini, F.: Robust synthesis of feedforward compensators. IEEE Transactions on Automatic Control 44(8), 1578–1582 (1999)

53. Grosdidier, P., Morari, M.: Interaction measure for systems under decentralized control. Automatica 22(3), 309–319 (1986)

54. Gross, E., Tomizuka, M., Messner, W.: Cancellation of discrete-time unstable zeros by feedforward control. Transactions of the ASME: Journal of Dynamic Systems, Measurement and Control 116(1), 33–38 (1994)

55. Guo, L.: Reducing the manufacturing costs associated with hard disk drives – a new disturbance rejection control scheme. IEEE/ASME Transactions on Mechantronics 2(2), 77–85 (1997)

56. Güvenç, B.A., Güvenç, L.: Robustness of disturbance observers in the presence of structured real parametric uncertainty. In: Proc. of the 2001 American Control Conference, Arlington, VA, USA, June 2001, pp. 4222–4227 (2001)

57. Güvenç, L.: Stability and performance robustness analysis of repeptitive control systems using structured singular values. Transactions of the ASME: Journal of Dynamic Systems, Measurement, and Control 118(3), 593–597 (1996)

58. Haack, B., Tomizuka, M.: The effect of adding zeros to feedforward controllers. Transactions of the ASME: Journal of Dynamic Systems, Measurement and Control 113(1), 6–10 (1991)

59. Hara, S., Omata, T., Nakano, M.: Synthesis of repetitive control systems and their application. In: Proc. of the 24th IEEE Conference on Decision and Control, Fort Lauderdale, FL, USA, December 1985, pp. 1387–1392 (1985)

60. Hara, S., Yamamoto, Y.: Stability of repetitive control systems. In: Proc. of the 24th IEEE Conference on Decision and Control, Fort Lauderdale, FL, USA, December 1985, pp. 326–327 (1985)

61. Hara, S., Yamamoto, Y., Omata, T., Nakano, M.: Repetitive control system: A new type servo system for periodic exogenous signals. IEEE Transactions on Automatic Control 33(7), 659–668 (1988)

62. Hassibi, B., Sayed, A.H., Kailath, T.: Indefinite-Quadratic Estimation and Control. A Unified Approach to $\mathcal{H}_2$ and $\mathcal{H}_\infty$ Theories. Society for Industrial and Applied Mathematics (1999)

63. Heuberger, P.S.C., Van den Hof, P.M.J., Wahlberg, B.: Modelling and Identification with Rational Orthogonal Basis Functions. Springer, London (2005)

64. Hillerström, G., Sternby, J.: Application of repetitive control to a peristaltic pump. Transactions of the ASME: Journal of Dynamic Systems, Measurement and Control 116(4), 786–789 (1994)

65. Hillerström, G., Sternby, J.: Repetitive control using low order models. In: Proc. of the 1994 American Control Conference, Baltimore, MD, USA, June 1994, pp. 1873–1878 (1994)

66. Hillerström, G., Walgama, K.: Repetitive control theory and applications - a survey. In: Proc. of the 13th IFAC World Congress, San Francisco, CA, USA, June-July 1996, pp. 1–6 (1996)

67. Horikawa, O., Sato, K., Shimokohbe, A.: An active air journal bearing. Nanotechnology 3(2), 84–90 (1992)

68. Horikawa, O., Shimokohbe, A.: An active air bearing. JSME International Journal, Series 3 33(1), 55–60 (1990)

69. Horowitz, I.M.: Synthesis of Feedback Systems. Academic Press, New York (1963)

70. Hostetter, G.H., Meditch, J.S.: On the generalization of observers to systems with unmeasurable, unknown inputs. Automatica 9(6), 721–724 (1973)

71. Hozumi, J., Hara, S., Fujioka, H.: Robust servo problem with $\mathcal{H}_\infty$ norm constraint. International Journal of Control 66(6), 803–823 (1997)

72. Hu, T., Teel, A.R., Lin, Z.: Lyapunov characterization of forced oscillations. Automatica 41(10), 1723–1735 (2005)
73. Hunt, L.R., Meyer, G., Su, R.: Noncausal inverses for linear systems. IEEE Transactions on Automatic Control 41(4), 608–611 (1996)
74. Inoue, T.: Practical repetitive control system design. In: Proc. of the 29th IEEE Conference on Decision and Control, Honolulu, Hawaii, December 1990, pp. 1673–1678 (1990)
75. Inoue, T., Nakano, M., Iwai, S.: High accuracy control of servomechanism for repeated contouring. In: Proc. of the 10th Annual Symposium on Incremental Motion, Control System and Devices, pp. 285–292 (1981)
76. Inoue, T., Nakano, M., Kubo, T., Matsumoto, S., Baba, H.: High accuracy control of a proton synchrotron magnet power supply. In: Proc. of the 8th IFAC World Congress, Kyoto, Japan, August 1981, pp. 3137–3142 (1981)
77. Iwasaki, T., Hara, S.: Generalized KYP lemma: unified frequency domain inequalities with design applications. IEEE Transactions on Automatic Control 50(1), 41–59 (2005)
78. Iwasaki, T., Skelton, R.E.: All controllers for the general $\mathcal{H}_\infty$ control problem: LMI existence conditions and state-space formulas. Automatica 30(8), 1307–1317 (1994)
79. Kalman, R.E.: Lyapunov functions for the problem of lur'e in automatic control. Proc. of the National Academy of Sciences of the United States of America 49(2), 201–205 (1963)
80. Kasac, J., Novakovic, B., Majetic, D., Brezak, D.: Passive finite-dimensional repetitive control of robot manipulators. IEEE Transactions on Control Systems Technology 16(3), 570–576 (2008)
81. Kempf, C., Messner, W., Tomizuka, M., Horowitz, R.: Comparison of four discrete-time repetitive control algorithms. IEEE Control Systems Magazine 13(6), 48–54 (1993)
82. Kim, B.K., Chung, W.K.: Advanced disturbance observer design for mechanical positioning systems. IEEE Transactions on Industrial Electronics 50(6), 1207–1216 (2003)
83. Kim, B.S., Li, J., Tsao, T.C.: Two-parameter robust repetitive control with application to a novel dual-stage actuator for noncircular machining. IEEE/ASME Transactions on Mechatronics 9(4), 644–652 (2004)
84. Kim, D.H., Tsao, T.C.: Robust performance control of electrohydraulic actuators for electronic cam motion generation. IEEE Transactions on Control System Technology 8(2), 220–227 (2000)
85. Kollar, I.: Frequency Domain System Identificaton Toolbox for Use with Matlab. The Mathworks Inc., Natic (2007)
86. Köroğlu, H., Morgül, Ö.: Discrete-time LQ optimal repetitive control. In: Proc. of the 1999 American Control Conference, San Diego, CA, USA, June 1999, pp. 3287–3291 (1999)
87. Köroğlu, H., Morgül, Ö.: Time-varying repetitive control for better transient response and stochastic behaviour. Electronics Letters 37(17), 1101–1102 (2001)
88. Köroğlu, H., Scherer, C.W.: An LMI approach to $\mathcal{H}_\infty$ synthesis subject to almost asymptotic regulation contraints. Systems & Control Letters 57(4), 300–308 (2008)
89. Köse, I.E., Scherer, C.W.: Robust feedforward control of uncertain systems using dynamic ICQs. In: Proc. of the 46th IEEE Conference on Decision and Control, New Orleans, LA, USA, December 2007, pp. 2181–2186 (2007)
90. Kučera, V.: Discrete Linear Control: The Polynomial Equation Approach. John Wiley and Sons, Inc, New York (1979)
91. Lee, R.C.H., Smith, M.C.: Robustness and trade-offs in repetitive control. Automatica 34(7), 889–896 (1998)

92. Li, J., Tsao, T.C.: Robust performance repetitive control systems. Transactions of the ASME: Journal of Dynamic Systems, Measurement, and Control 123(3), 330–337 (2001)

93. Lindquist, A., Yakubovich, V.A.: Universal regulators for optimal tracking in discrete-time systems affected by harmonic disturbances. IEEE Transactions on Automatic Control 44(9), 1688–1704 (1999)

94. Liu, Z., Vandenberghe, L.: Low-rank structure in semidefinite programs derived from the KYP lemma. In: Proc. of the 46th IEEE Conference on Decision and Control, New Orleans, LA, USA, December 2007, pp. 5652–5659 (2007)

95. Lu, Y., Messner, W.C.: Disturbance observer design for tape transport control. In: Proc. of the 2001 American Control Conference, Arlington, VA, USA, June 2001, pp. 2567–2571 (2001)

96. Luenberger, D.: Observers for multivariable systems. IEEE Transactions on Automatic Control 11(2), 190–197 (1966)

97. Maciejowski, J.M.: Multivariable Feedback Design. Addison-Wesley, Wokingham (1989)

98. Manayathara, T.J., Tsao, T.C., Bentsman, J., Ross, D.: Rejection of unknown periodic load disturbances in continuous steel casting process using learning repetitive control approach. IEEE Transactions on Control Systems Technology 4(3), 259–265 (1996)

99. Mattavelli, P., Tubiana, L., Zigliotto, M.: Torque-ripple reduction in PM synchronous motor drives using repetitive current control. IEEE Transactions on Power Electronics 20(6), 1423–1431 (2005)

100. Medvedev, A., Hillerström, G.: On perfect disturbance rejection. In: Proc. of the 32nd IEEE Conference on Decision and Control, San Antonio, TX, USA, December 1993, pp. 1324–1329 (1993)

101. Moon, J.H., Lee, M.N., Chung, M.J.: Repetitive control for the track-following servo system of an optical disk drive. IEEE Transactions on Control Systems Technology 6(5), 663–670 (1998)

102. Morari, M., Zafirou, E.: Robust Process Control. Prentice-Hall, Englewood Cliffs (1989)

103. Ohnishi, K.: A new servo method in mechatronics. Transactions of the Japanese Society of Electrical Engineers 107-D, 83–86 (1987)

104. Ohnishi, K.: Realization of fine motion control based on disturbance observer. In: Proc. of the 10th International Workshop on Advanced Motion Control, Trento, Italy, March 2008, pp. 1–8 (2008)

105. Osburn, A.W., Franchek, M.A.: Designing robust repetitive controllers. Transactions of the ASME: Journal of Dynamic Systems, Measurement, and Control 126(4), 865–872 (2004)

106. Packard, A., Doyle, J.: The complex structured singular value. Automatica 29(1), 71–109 (1993)

107. Papoulis, A.: Signal Analysis. McGraw-Hill, New York (1977)

108. Peery, T.E., Özbay, H.: $\mathcal{H}_\infty$ optimal repetitive controller design for stable plants. Transactions of the ASME: Journal of Dynamic Systems, Measurement, and Control 119(3), 541–547 (1997)

109. Pintelon, R., Schoukens, J.: System Identification: A frequency Domain Approach. Institute of Electrical and Electronics Engineers, Inc., New York (2001)

110. Pipeleers, G., Demeulenaere, B., Al-Bender, F., De Schutter, J., Swevers, J.: Optimal performance trade-offs in repetitive control: experimental validation on an active air bearing setup. IEEE Transactions on Control Systems Technology 17(4), 970–979 (2009)

111. Pipeleers, G., Demeulenaere, B., De Schutter, J., Swevers, J.: Generalized repetitive control: relaxing the one-period-delay structure. IET Control Theory and Applications (in press, 2009)

112. Pipeleers, G., Demeulenaere, B., Swevers, J., De Schutter, J.: Robust high-order repetitive control: optimal performance trade-offs. Automatica 44(10), 2628–2634 (2007)

113. Popov, V.M.: Absolute stability of nonlinear systems of automatic control. Automation and Remote Control 22, 857–875 (1962); Russian original in August 1961

114. Randke, A., Gao, Z.: A survey of state and disturbance obcervers for practioners. In: Proc. of the 2006 Amrerican Control Conference, Minneapolis, MN, USA, June 2006, pp. 5183–5188 (2006)

115. Ringey, B.P., Pao, L.Y., Lawrence, D.A.: Nonminimum phase dynamic inversion for settle time applications (2008); reprint accepted for publication in IEEE Transactions on Control Systems Technology

116. Ryoo, J.R., Jin, K.B., Moon, J.H., Chung, M.J.: Track-following control using a disturbance observer with asymptotic disturbance rejection in high-speed optical disk drives. IEEE Transactions on Consumer Electronics 49(4), 1178–1185 (2003)

117. Saberi, A., Stoorvogel, A.A., Sannuti, P.: Control of Linear Systems with Regulation and Input Contraints. Springer, London (2000)

118. Saberi, A., Stoorvogel, A.A., Sannuti, P., Shi, G.: On optimal output regulation for linear systems. International Journal of Control 76(4), 319–333 (2003)

119. Sadech, N., Horowitz, R., Kao, W.W., Tomizuka, M.: A unified approach to the design of adaptive and repetitive controllers for robotic manipulators. Transactions of the ASME: Journal of Dynamic Systems, Measurement and Control 112(4), 618–629 (1990)

120. Sayed, A.H., Kailath, T.: A survey of spectral factorization methods. Numerical Linear Algebra with Applications 8(6-7), 467–496 (2001)

121. Scherer, C., Gahinet, P., Chilali, M.: Multiobjective output-feedback control via LMI optimization. IEEE Transactions on Automatic Control 42(7), 896–911 (1997)

122. Scherer, C., Weiland, S.: Linear Matrix Inequalities in Control. Lecture Notes for a course of the Dutch Institute of Systems and Control (2005)

123. Scherer, C.W.: LMI relaxations in robust control. European Journal of Control 12(1), 3–29 (2006)

124. Schoukens, J., Rolain, Y., Simon, G., Pintelon, R.: Fully automated spectral analysis of periodic signals. IEEE Transactions on Instrumentation and Measurement 52(4), 1021–1024 (2003)

125. Schrijver, E., van Dijk, J.: Disturbance observers for rigid mechanical systems: equivalence, stability, and design. Transactions of the ASME: Journal of Dynamic Systems, Measurement, and Control 124(4), 539–548 (2002)

126. Scorletti, G., Fromion, V.: Further results on the design of robust $\mathscr{H}_\infty$ feedforward controllers. In: Proc. of the 45th IEEE Conference on Decision and Control, San Diego, CA, USA, December 2006, pp. 3560–3565 (2006)

127. Serrani, A.: Rejection of harmonic disturbances at the controller input via hybrid adaptive external models. Automatica 42(11), 1977–1985 (2006)

128. Shaw, F.R., Srinivasan, K.: Discrete-time repetitive control system design using the regeneration spectrum. Transactions of the ASME: Journal of Dynamic Systems, Measurement, and Control 115(2A), 228–237 (1993)

129. Shim, H., Jo, N.H., Son, Y.I.: A new disturbance observer for non-minimum phase linear systems. In: Proc. of the 2008 American Control Conference, Seattle, WA, USA, June 2008, pp. 3385–3389 (2008)

130. Shim, H., Joo, Y.J.: State space analysis of disturbance observer and a robust stability condition. In: Proc. of the 46th IEEE Conference on Decision and Control, New Orleans, LA, USA, December 2007, pp. 2193–2198 (2007)
131. Skogestad, S., Postlethwaite, I.: Multivariable Feedback Control - Analysis and Design. John Wiley & Sons, Chichester (2005)
132. Smith, C., Tomizuka, M.: A cost effective repetitive controller and its design. In: Proc. of the 2000 American Control Conference, Chicago, IL, USA, June 2000, pp. 1169–1174 (2000)
133. Sparks, A.G., Bernstein, D.S.: Asymptotic regulation with $\mathcal{H}_2$ disturbance rejection. In: Proc. of the 33rd IEEE Conference on Decision and Control, Lake Buena Vista, FL, USA, December 1994, pp. 3614–3615 (1994)
134. Srinivasan, K., Shaw, F.R.: Analysis and design of repetitive control systems using the regeneration spectrum. Transactions of the ASME: Journal of Dynamic Systems, Measurement, and Control 113(2), 216–222 (1991)
135. Steinbuch, M.: Repetitive control for systems with uncertain period-time. Automatica 38(12), 2103–2109 (2002)
136. Steinbuch, M., Weiland, S., Singh, T.: Design of noise and period-time robust high order repetitive control, with application to optical storage. Automatica 43(12), 2086–2095 (2007)
137. Stoorvogel, A.A., Saberi, A., Sannuti, P.: Performance with regulation constraints. Automatica 36(10), 1443–1456 (2000)
138. Sung, H.K., Hara, S.: Properties of sensitivity and complementary sensitivity functions in single-input single-output digital control systems. International Journal of Control 48(6), 2429–2439 (1988)
139. Tammi, K., Hatonen, J., Daley, S.: Novel adaptive repetitive algorithm for active vibration control of a variable-speed rotor. Journal of Mechanical Science and Technology 21(6), 855–859 (2007)
140. Tesfaye, A., Lee, H.S., Tomizuka, M.: A sensitivity optimization approach to design of a disturbance observer in digital motion control systems. IEEE/ASME Transactions on Mechatronics 5(1), 32–38 (2000)
141. Toh, K.C., Todd, M.J., Tütüncü, R.H.: SDPT3 - a matlab software package for semidefinite programming. Optimization Methods and Software 11, 545–581 (1999)
142. Tomizuka, M.: Zero phase error tracking algorithm for digital control. Transactions of the ASME: Journal of Dynamic Systems, Measurement and Control 109(1), 65–68 (1987)
143. Tomizuka, M., Chen, M.S., Renn, S., Tsao, T.C.: Tool positioning for noncircular cutting with lathe. Transactions of the ASME: Journal of Dynamic Systems, Measurement and Control 109(2), 176–179 (1987)
144. Tomizuka, M., Chew, K.K., Yang, W.C.: Disturbance rejection through an external model. Transactions of the ASME: Journal of Dynamic Systems, Measurement, and Control 112(4), 559–564 (1990)
145. Tomizuka, M., Tsao, T.C., Chew, K.K.: Discrete-time domain analysis and synthesis of repetitive controllers. In: Proc. of the 1988 American Control Conference, Atlanta, GA, USA, June 1988, pp. 860–866 (1988)
146. Tomizuka, M., Tsao, T.C., Chew, K.K.: Analysis and synthesis of discrete-time repetitive controllers. Transactions of the ASME: Journal of Dynamic Systems, Measurement, and Control 111(3), 353–358 (1989)
147. Torfs, D., De Schutter, J., Swevers, J.: Extended bandwidth zero phase error tracking control of nonminimal phase systems. Transactions of the ASME: Journal of Dynamic Systems, Measurement and Control 114(3), 347–351 (1992)

148. Tsao, T.C., Tomizuka, M.: Robust adaptive and repetitive digital tracking control and application to a hydraulic servo for noncircular machining. Transactions of the ASME: Journal of Dynamic Systems, Measurement and Control 116(1), 24–32 (1994)
149. Tütüncü, R.H., Toh, K.C., Todd, M.J.: Solving semidefinite-quadratic-linear programs using SDPT3. Mathematical Programming 95(2), 189–217 (2003)
150. Vandenberghe, L., Balakrishnan, V.R., Wallin, R., Hansson, A., Roh, T.: Interior-point algorithms for semidefinite programming problems derived from the KYP lemma. In: Henrion, D., Garulli, A. (eds.) Positive Polynomials in Control. LNCIS, pp. 195–238. Springer, Heidelberg (2005)
151. Walgama, K.: On the Control of Systems with Input Saturation or Periodic Disturbances. PhD thesis, Luleå University of Technology (1991)
152. Walgama, K.S., Sternby, J.: A feedforward controller design for periodic signals in non-minimum phase processes. International Journal of Control 61(3), 695–718 (1995)
153. Wang, C.C., Tomizuka, M.: Design of robustly stable disturbance observers based on closed-loop consideration using $\mathcal{H}_\infty$ optimization and its application to motion control systems. In: Proc. of the 2004 American Control Conference, Boston, MA, USA, June-July 2004, pp. 3764–3769 (2004)
154. Wang, Q.G., Zhang, Y., Huang, X.G.: Virtual feedforward control for asymptotic rejection of periodic disturbance. IEEE Transactions on Industrial Electronics 49(3), 566–573 (2002)
155. White, M.T., Tominuka, M., Smith, C.: Improved track-following in magnetic disk drives using a disturbance observer. IEEE/ASME Transactions on Mechatronics 5(1), 3–11 (2000)
156. Wu, Q., Saif, M.: Repetitive learning observer based actuator fault detection, isolation, and estimation with application to a satellite attitude control system. In: Proc. of the 2007 American Control Conference, New York, NY, USA, July 2007, pp. 251–256 (2007)
157. Yakubovich, V.A.: Solution of certain matrix inequalities encountered in non-linear regulation theory. Doklady Akademii Nauk SSSR 143, 1304–1307 (1962); English translation in Soviet mathematics - Doklady 3, 620–623 (1962)
158. Yang, W.C., Tomizuka, M.: Disturbance rejection through an external model for non-minimum phase systems. Transactions of the ASME: Journal of Dynamic Systems, Measurement, and Control 116(1), 39–44 (1994)
159. Youla, D.C., Bongiorno, J.J., Jabr, H.A.: Modern Wiener-Hopf design of optimal controllers, part 1: The single-input-output case. IEEE Transactions on Automatic Control 21(1), 3–13 (1976)
160. Youla, D.C., Jabr, H.A., Bongiorno, J.J.: Modern Wiener-Hopf design of optimal controllers, part 2: The multivariable case. IEEE Transactions on Automatic Control 21(3), 319–338 (1976)
161. Zhang, X.Y., Shinshi, T., Li, L.C., Choi, K.B., Shimokohbe, A.: Precision control of radial magnetic bearing. In: Proc. of the 10th International Conference on Precision Engineering, Yokohama, Japan, July 2001, pp. 714–718 (2001)
162. Zhang, X.Y., Shinshi, T., Li, L.C., Shimokohbe, A.: A combined repetitive control for precision rotation of magnetic bearing. Precision Engineering 27(3), 273–282 (2003)
163. Zhou, K., Doyle, J.C., Glover, K.: Robust and Optimal Control. Prentice-Hall, Upper Saddle River, NJ (1996)
164. Zhou, K.L., Wang, D.W.: Digital repetitive learning controller for three-phase CVCF PWM inverter. IEEE Transactions on Industrial Electronics 48(4), 820–830 (2001)
165. Zhou, K.L., Wang, D.W., Low, K.S.: Periodic errors elimination in CVCF PWM DC/AC converter systems: repetitive control approach. IEE Proceedings - Control Theory and Applications 147(6), 694–700 (2000)

166. Zou, Q.Z.: Optimal preview-based stable-inversion for output tracking of nonminimum-phase linear systems. In: Proc. of the 46th IEEE Conference on Desicion and Control, New Orleans, LA, USA, December 2007, pp. 5258–5263 (2007)
167. Zou, Q.Z., Devasia, S.: Preview-based stable-inversion for output tracking of linear systems. Transactions of the ASME: Journal of Dynamic Systems, Measurement and Control 121(4), 625–630 (1999)
168. Zou, Q.Z., Devasia, S.: Preview-based optimal inversion for output tracking: application to scanning tunneling microscope. IEEE Transactions on Control Systems Technology 12(3), 375–386 (2004)

Index

Lecture Notes in Control and Information Sciences

Edited by M. Thoma, F. Allgöwer, M. Morari

Further volumes of this series can be found on our homepage:
springer.com